MAN
OVER
MACHINE

AF650149

MAN OVER MACHINE

A Path towards Peace

BHARAT DOGRA

Vitasta
LET KNOWLEDGE SPREAD

Published by
Renu Kaul Verma
Vitasta Publishing Pvt Ltd
2/15, Ansari Road, Daryaganj
New Delhi-110 002
info@vitastapublishing.com

ISBN 978-93-86473-75-2
© Bharat Dogra
First Edition 2020
MRP ₹495

All Rights Reserved.
No part of this publication may be reproduced, stored in a retrieval system, or transmitted in any form, or by any means—electronic, mechanical, photocopying, recording or otherwise—without the prior permission of the publisher. Opinions expressed in this book are the author's own.

Cover and layout by Somesh Kumar Mishra
Printed by Chaman Enterprises

This book is for granddaughter
Srishti Priya and her friends
in the hope that they grow
up in a safer and more peaceful world

I've said time and again - the truly valiant are those who resist by non-violent means.

—Mahatma Gandhi

World View

We may all be happy and grateful that destiny gifted us with such an enlightened contemporary, (Gandhi) a role model for generations to come. Generations to come will scarce believe that such a one as this walked the earth in flesh and blood.

—Albert Einstein

Christ gave us the goals and Mahatma Gandhi the tactics.

—Martin Luther King

Contents

Introduction

There are several ways of looking at the legacy of Mahatma Gandhi. My effort in this book is to concentrate more on how this can be most useful to our troubled world now and in the years to come.

This is a world burdened by most serious environmental problems and threatened by most destructive weapons. The activist poet Wendell Berry asserts: The increasingly aggressive trends in world economy are tied up inseparably with massive plunder of others resources, hence with conquest and war. So, let us look at Mahatma Gandhi's catchwords: Meet the needs of all and greed of none. That – one's life should be lived sustainably.

In this contemporary context we examine the enduring Gandhian message of peace, non-violence, peaceful resistance of anything unjust, as well as a simple, low consumption yet happy and fulfilling life, based on need and not greed.

We need to understand first what this Gandhian perspective means for big policy changes as well as for individual and community life. What has been its impact on various social movements in more recent times?

On the one hand, our deeply troubled world needs big changes at the policy level and on the other hand, we need changes in the everday life and thinking of ordinary people to create grassroots support for badly-needed policy changes. In this book both aspects are discussed. We also need inspirational stories from real life and in this book we discuss several social movements inspired to a significant extent by Gandhiji. Then we also have stories of inspirational individuals who lived life much in tune with various ideals of Gandhiji. This brings us closer to what can be achieved at the practical level. I have taken care to include those individuals whom I knew well and those movements which I have covered several times as a journalist.

This book is based on very high regard for Gandhiji's efforts for peaceful resistance to injustice and their great contribution to the freedom movement of India. This remains a matter of great relevance to present times, even though direct colonial rule ended a long time ago. Despite this great regard, however, we also need to hear voices which give a different viewpoint – it would be un-Gandhian not to do so! This is particularly more necessary when spreading the message of Gandhiji among youth.

If youth have any doubts then these should be settled through open discussion and any critical facts which emerge should also be discussed in a free and frank way. This is why in this book, which is extremely respectful towards the great contributions of Gandhiji, we have kept the door open for viewpoints which examine the role of Gandhiji at certain junctures of history in critical ways.

No one who reached out to so many people with continuity for so many years could be expected to *always* say the right thing and make exactly the right choice among the various options open to him. No one could be or can be expected to be one hundred per cent free from the myths and prejudices of his/our times and the social classes in which we are brought up. Hence, it will not be fair to reject any great work of enduring importance just because of occasional careless statements or misjudgments.

We take this view so that while being critical where we feel this is needed, we remain firmly on our main track of examining the enduring, in fact, increasing significance of the work and ideas of Mahatma Gandhi for our times and for future.

To do this, let us first examine the most urgent issues of concern in the second decade of the 21st century, at a time when a quarter of the new century is almost over and we have to address a very young population, averagely 25 years old with a life expectancy of 75 years, may be travelling to the moon and outer space frequently – at the same time, without a say in limiting wars and territorial conflicts, unable to control commercialisation and commodification, and unable to compromise with the environment.

So far, this young world population is only taking what is being given by the older generation in accordance to what this older generation thinks is right. This needs to change.

To this young generation of the world, we bring a book that discusses wars and what perhaps Gandhi would have said on this; on economics in times of climate change; how professionals can be and should be activists and steer sustainable life on earth.

SUSTAINABLE DEVELOPMENT AND ENVIRONMENT PROTECTION

The true measure of any society can be found in how it treats its most vulnerable members.

—Mahatma Gandhi

Many of his (Gandhi's) principles have universal application and validity.

—U Thant

In Times of Climate Change
Gandhian Economics more Relevant

WHILE MAHATMA GANDHI'S role as a freedom fighter and social reformer has been widely recognised, his contribution as an ecologist who foresaw with remarkable foresight the environmental crisis of the coming decades has been generally neglected. Of course, the word *ecology* was not much in vogue then, yet Gandhi's vision of a better world includes a very clear understanding of what needs to be done for the protection of environment.

In this context, two key statements of Gandhi need to be stressed. The first statement provides a guideline on how development choices should be made. The Mahatma told policy-makers and others that whenever they are in doubt 'recall the face of the poorest and the weakest man whom you may have seen and ask yourself if the step you contemplate is going to be of any use to him? Will it restore him to control over his own life and destiny?' In other words, the needs of the poorest people should receive the topmost priority in development planning.

The second key statement of Mahatma Gandhi tells us that while the basic needs of all people should be met, there should also be consciousness regarding placing a limit on consumption. Although the pressures on nature had not become so acute in Gandhi's lifetime and there were no scare either of ozone depletion or climate change, this wise man had the foresight to see that our planet's carrying capacity is not infinite and some restraint will have to be placed on how much pressure we can place on its air, water, forests and minerals

Mahatma Gandhi said, 'Nature has enough to satisfy the needs of everyone but nothing to satisfy the greed of even a few'.

Due to the unequal distribution of power, it is possible for a relatively small number of people to corner an extremely large share of the earth's resources. The United Nations Development Programme in a report said in 1996, the net worth of 358 richest people of the world is equal to the combined income of the poorest 45% of the world's population – 2.3 billion people. The richest 20% of the world's people corner 85% of the world's income.

People in industrial countries consume nearly nine times as much commercial energy per capita as people in developing countries, though they constitute only a fifth of the world's population. It is, therefore, clear that the pressures on environment are largely a result of the excessive consumption of the richest 20 per cent of world's people.

Mahatma Gandhi stated quite clearly that the greed of even a few people, their tendency to pursue a highly affluent and wasteful life-style, can prove very destructive to nature. This is evident today in the global warming crisis, very high levels of air and water pollution and various other alarming aspects of ecological ruin. The Gandhian solution to the environment crisis is clear – curb this greed of the few.

If we can convince the richest people of the desirability of not increasing their consumption and accumulation beyond a point, then this will reduce the pressure on nature and help us to protect environment. In addition, it will become easier to meet the needs of the poor.

Need for an Environment Space

In view of the enormity of the environment crisis, it is important today to speak of an 'environment space' within which economic and industrial activity should be contained if it is not to become destructive. When the richest people vacate some of this space by reducing their consumption, then it becomes easier to meet the basic needs of the poorest people.

The two most important challenges of our time are protecting environment and meeting the basic needs of all. The Gandhian response to both challenges is simple and similar – release resources form the grip of the very rich people so that the needs of the poor can be met.

The life-style of the richest people is attractive and so it soon becomes a model for others. Mahatma Gandhi said clearly that this is a model not worth emulating because it is destructive to nature. Instead, he tried throughout his life to experiment with low-cost food, farming, education and medicare which could meet the needs of all people.

Mahatma Gandhi did not have access to the extensive data on ozone depletion, acid rain, biodiversity loss and climate change which is available to scientists and scholars today. But his quest for deeper truth of life had taken him to a higher state of understanding from where he could anticipate some of the greatest threats being faced by humanity.

Gandhi's understanding of the ecological crisis was much ahead of his times and is still of great relevance today in solving some of humankind's most pressing problems.

Mahatma Gandhi challenged well-entrenched concepts of what is development – a task which must have been even more difficult then when some of the worst environmental aspects of this development phase had not been manifested yet. In 1925 he wrote:

> What is industrialism, but a control of the majority by a small minority? There is nothing attractive about it, nor is there anything inevitable in it. If the majority simply wills to say 'no' to the blandishments of the minority, the latter is powerless for mischief. It is good to have faith in human nature. What may be hoped for is that Europe – on account of her fine and scientific intellect – will realise the obvious and retrace her steps, and that from the demoralising industrialism she will find a way out. It will not necessarily be a return to the old absolute simplicity. But it will have to be a reorganisation in which village life will predominate, and in which brute and material force will be subordinated to the spiritual force.

In this statement it is evident that

- 1. He considered the existing industrialisation pattern to be anti-democratic and against the interest of most people.
- 2. He firmly believed that alternatives were available.
- 3. These alternatives will have greater role for village community life and spiritual development, he thought.

Further, he said in the same year:

> We know, too, that each nation has its own characteristics and individuality. I claim that to industrialise India in the same sense as Europe is to attempt the impossible.

In 1929, he repeated optimistically that the West will itself realise the futility of this path, but in any case, India should never pursue this path. He wrote:

> There is a growing body of enlightened opinion in the West which distrusts this civilisation which has insatiable material ambition at one end and consequent war at the other. But whether good or bad, why must India become industrial in the Western sense?

While thinking of an alternative path of development, one thing Gandhi was very clear about was that it will not be based on exploitation. He wrote in 1929: Surely exploitation means usurpation. And usurpation can never be reconciled with spiritualism.

Once Gandhi was asked whether he will like India to develop as much as Britain. Gandhi replied that Britain was such a small country but it required the plunder of half the planet to give it this development, so for a big country like India to develop in the same way will probably require the plunder of several planets. But he was certain that even if these planets were available, he never would want his country to follow this path. He wrote in 1940:

> I have no idea of exploiting other countries for the benefit of India. We are suffering from the poisonous disease of exploitation ourselves, and I would not like my country to be guilty of any such thing.

He went a step further and asked all rich persons to self examine how their wealth has come – directly or indirectly from the exploitation of poor people. As a penance, he asked them to do some physical work which is also related to the welfare of the poor (such as spinning yarn for cottage-scale cloth industry). This view is evident in a question-answer session in 1924:

> Question: Why should I, who have no need to work for food, spin?
>
> Answer: Because I am eating what does not belong to me. I am living on the spoilation of my countrymen. Trace the course of every pice (paisa) that finds its way into your pocket, and you will realise the truth of what I write.

One aspect of this exploitation which particularly pained him was the exploitation of villages by cities, of rural life by urban life. He wrote in 1927, 'The half-a-dozen modern cities are an excrescence and serve, at the present moment, the evil purpose of draining the life-blood of the villages'.

In 1936, he wrote in more specific terms:

> Little flour mills are ousting the chakki, oil mills the village ghani, rice mills the village dhenki, sugar mills the village gud-pans, etc. This displacement of village labour is impoverishing the villagers and enriching the monied men. If the process continues sufficiently long, the villages will be destroyed without any further effort.

In this system of exploitation, a particularly destructive role was played by labour-displacing machinery. He wrote in 1936:

> A factory employs a few hundreds and renders thousands unemployed. I may produce tons of oil from an oil mill, but I also drive thousands of oilmen out of employment. I call this destructive energy, whereas production by the labour of millions of hands is constructive and conductive to the common good. Mass production through power-driven machinery, even when state-owned, will be of no avail.

Earlier in 1931 he had warned:

> If the craze for the machinery method continues, it is highly likely that a time will come when we shall be so incapacitated and weak that we shall begin to curse ourselves for having forgotten the use of the living machines given to us by God.

However, he was clearly not against all machinery, as is evident from this question answer session in 1924.

> Question: Are you against all machinery?
>
> Answer: How can I be when I know that even this body is a most delicate piece of machinery? The spinning wheel itself is a machine; a little tooth-pick is a machine. What I object to is the craze for machinery, not machinery as such. The craze is for what they call 'labour-saving' machinery. Men go on saving labour, till thousands are without work and thrown on the open streets to die of starvation.

When asked what kind of machinery he approved, Gandhi said in 1935:

> Any machinery which does not deprive masses of men of the opportunity to labour, but which helps the individual and adds to his efficiency, and which a man can handle at will without being its slave.

Giving a balanced view many years ago, Gandhi summarised his view on machinery:

> The heavy machinery for work of pubic utility which cannot be undertaken by human labour has its inevitable place, but all that would be owned by the State and used entirely for the benefit of the people. I can have no consideration for machinery which is meant to enrich the few at the expense of the many or without cause to displace the useful labour of many. Take printing presses.

> They will go on. Take surgical instruments. How can one make them with one's hands?

Mahatma Gandhi's views on machinery did not remain confined to a theoretical level. Perhaps more than anywhere else in the world, these views found practical application as these became an integral part of the freedom movement of India.

Swaraj from the Spinning Wheel

An interesting and encouraging aspect of the freedom movement in India was that along with the struggle against colonial rule, vigorous efforts were made to find an alternative path of development. While several people in India were eager to 'develop' as much as the British and later some others wanted to industrialise as rapidly as the Soviets, there were others who kept alive the concept of small and cottage-scale development to be based in largely self-reliant rural communities.

This viewpoint was most vigorously articulated by Mahatma Gandhi who popularised the spinning wheel or 'charkha' to symbolise this aspect of self-rule or 'swaraj'. Gandhi's early experiences with the charkha are still significant in the context of the 'large vs small, global vs local' debate.

In a significant book, *Hind Swaraj,* or *Indian Home Rule,* he wrote in 1908:

> It is machinery that has impoverished India. It is difficult to measure the harm that Manchester has done to us. It is due to Manchester that Indian handicraft has all but disappeared. But I make a mistake. How can Manchester be blamed? We wore Manchester cloth and this is why Manchester wove it.

So, Gandhi argued strongly in favour of going back to self-reliant production of clothes in villages – a task in which charkha will have the crucial role of spinning the yarn, which will be used further by the handloom weavers to produce entirely hand-made cloth, called khadi or khaddar.

Mahatma Gandhi was frequently asked why he gave so much importance to the spinning wheel. He replied in 1923: Economic inquiries conducted between 1798 and 1814 show how many hundreds of thousands of our men, women and children worked on this (spinning) industry – mostly in their leisure time – each day and earned crores of rupees annually (1 crore = 10 million).

In 1931 he described his thought-process more clearly:

> 'What is the kind of service' I asked myself, 'that the teeming millions of the India most need at the present time, that can be easily understood and appreciated by all, that is easy to perform and will at the same time enable the crores of our semi-starved countrymen to live?' And the reply came that it is the universalisation of Khadi or the spinning wheel alone that can fulfil these conditions.

The spinning wheel in Indian conditions was the best example of appropriate technology, because, as Gandhi said in 1927: It uses machinery for the service of the poorest in their own cottages. Further, by taking essential industrial activity to village cottages 'Khaddar is an attempt to revise and reverse the process and establish a better relationship between the cities and the villages'.

Mahatma Gandhi recognised that given the low per capita land availability in villages, the Indian peasants needed some additional craft work that could be pursued easily by the family without much capital investment. He wrote quite clearly in

1919: Without a cottage industry the Indian peasant is doomed. He cannot maintain himself from the produce of the land.

In 1921 he wrote, 'I have seen women beaming with joy to see the spinning wheel work, for they know that they can through that rustic instrument both feed and clothe themselves'.

An interesting thing was that while Gandhi was arguing so strongly in favour of the charkha, he had not yet seen it. Mill-made cloth and yarn, Indian and foreign, had spread so rapidly that it was not easy to find a charkha. It was relatively easier to find handloom weavers, but they also used yarn which had been manufactured in mills. When Gandhi and his friends established a *satyagraha ashram* in Sabarmati (Gujarat), it was easier for them to install handlooms, or to get their cloth woven by handloom weavers, but it proved much more difficult to get hand-spun yarn. So, initially even the ashram had to remain dependent on the yarn supplied by the Indian mills. For Mahatma Gandhi, it was not enough to substitute British mills by Indian mills, he wanted to end the dependence on all mills. Gandhi wrote about this phase:

> We became impatient to be able to spin our own yarn. It was clear that, until we could do this ourselves, dependence on the mills would remain. We did not feel that we could render any service to the country by continuing as agents of Indian spinning mills.

But so complete had been the uprooting of the traditional system of cloth-making that it proved very difficult to re-establish its broken links. Till only a century or two ago, European travellers had themselves noted how cloth was being produced in almost all villages which were self-reliant in all stages including cultivation of cotton, spinning and weaving. But all these linkages had been broken up and even the type of cotton that was now grown was more suitable for spinning in mills. The problems of re-establishing the lost linkages proved particularly difficult in Gujarat (the province where Gandhi was staying), which had a large share of the early growth of Indian mills.

Gandhi wrote, 'No end of difficulties again faced us. We could get neither spinning wheel nor a spinner to teach us how to spin. The time passed on, and my impatience grew with time'.

Hope finally emerged in the form of a remarkable lady – Ganga Behn Majumdar – who promised to find the charkha and hand-spinners for Gandhi.

At last, after no end of wandering in Gujarat, Ganga Behn found the spinning wheel in Vijapur in the Baroda State. Quite a number of people there had spinning wheels in their homes, but had long since consigned them to the lofts as useless lumber. They expressed to Ganga Behn their readiness to resume spinning if someone promised to provide them with a regular supply of slivers, and to buy the yarn spun by them.

This was good news for Gandhi, but as slivers were still obtained from mills, some dependence on mills still remained. If one could use mill-slivers, why not use mill-yarn as well? So Gandhi again approached Ganga Behn to find carders who could supply slivers. She soon found a carder and trained a few youngsters to make slivers out of the carded cotton.

With the broken links thus re-established, the Satyagraha Asharm was soon able to make khadi (or cloth which is hand-spun as well as hand-woven).

When an Indian mill-owner heard of Gandhi's efforts, he called upon him to convince him that the best way of reducing dependence on imports was to establish more Indian mills.

'I am not doing exactly that', Gandhi replied, 'but I am engaged in the revival of the spinning wheel'.

'What is that?', the mill-owner asked, feeling still more at sea.

After explaining his work to him, Gandhi concluded, 'I swear by this form of Swadeshi, because through it I can provide work to the semi-starved semi-employed women of India. My idea is to get these women to spin yarn, and to clothe the people of India with khadi woven out of it'.

It is clear from this episode that Gandhi's concern was not confined to reducing the dependence on foreign mills, he was equally eager to reduce the villagers' dependence on domestic mills in the context of that produce which could be made by villagers themselves.

In 1924 he stated clearly, 'Just as we should not on any pretext whatsoever use foreign cloth, we should not use cloth manufactured in our mills'.

In fact, even much earlier, in 1908 he was clear that: It would be folly to assume that an Indian Rockfeller would be better than the American Rockfeller.

For Mahatma Gandhi, it was clear that it will be foolhardy for handloom weavers to rely on yarn supplied by mills, for mills see their interest in displacing them instead of helping them. In 1926 he wrote:

> It is wrong to rely upon an unfailing supply of mill yarn. Hand-weaving and mill-weaving are not complementary propositions. They are mutually antagonistic, the tendency of weaving mills, like all machinery, always being to displace the product of the hand. If, therefore, hand-weaving could become a supplementary industry on a large scale, it would have to be solely dependent on mills which would naturally squeeze the last pie from the weaver for the supply of yarn, and would scrap it at the

> first opportunity. On the other hand, hand-spinning and hand-weaving are mutually complementary. It is little known that a vast number of weavers of mill yarn are in the hands of sowcars (money lenders), and they must be, so long as they rely upon the mill product. The village economy demands that the weaver should receive his yarn not from the middlemen but from his fellow-worker, the farmer.

He said further that handloom weaving will progress automatically if the spinning wheel succeeds. It is bound to die if the wheel fails.

In 1927, in a prophetic message to the weavers of India be said:

> Every weaver who weaves yarn which is supplied by foreign mills or even by mills of India, places himself at the disposal of and at the tender mercy of the mills. You as weavers should realise that this hand-weaving which you are today controlling to a certain extent, will in time to come, slip away from your hands as soon as the mills of the world or the mills of India are ready to weave the patterns that you are today exclusively weaving. Let me inform you, if you do not know the fact already, that various able mill owners of the world are making experiments in order to weave the patterns which are today your monopoly. It is no fault of the mill owner or the mill industry that the mill industry is endeavouring day after day to take away the monopoly and take this trade in its own hands. To make continuous improvements in its machinery and to make continuous encroachment upon the handicrafts of the world is really the objective and the ideal of these great industrialists. Indeed, it is the condition of their very existence that they should try to take this trade also from off your hands.

In 1936, while evaluating the progress made by Khadi, he wrote with some satisfaction:

> The progress Khadi has made in terms of the millions, though little in itself, is comparatively the largest of all the other single industries. It distributes yearly the largest amount as wages among the largest number of wage-earners in the villages with the minimum of overhead charges, and every pice practically circulates among the people.

He was equally conscious of the shortcomings and regretted why most handloom owners still continued to use mill yarn.

He wrote: Similarly we should have made khadi workers weave also. It is only when we learn the art of weaving that we shall be able to understand fully the difficulties the weavers have yet to encounter in weaving the hand spun yarn, and shall be able to discover remedies for them.

Clearly, he was calling for innovation. On the one hand, he called for continuous technical improvement and on the other hand, he urged more and more people to adopt khadi clothes. In 1934 he wrote: Let this be borne in mind that khadi can be permanent only when it has obtained a permanent footing as a village wear.

Gandhi on Adam Smith

Gandhi's view on khadi present the key to his views on economics, or to what some scholars have called Gandhian Economics. He made it clear that his economic ideas were in some way the opposite of the views which were widely prevalent then, particularly that of Adam Smith.

He wrote in 1934:

> In reorganising your khadi production you should not forget that the science of khadi in some respects works on lines diametrically opposite to that of ordinary business.

> You know how Adam Smith in his *Wealth of Nations*, after laying down certain principles according to which economic phenomena are governed, went on to describe certain other things which constituted the 'disturbing factor' and prevented economic laws from having free play. Chief among these was the 'human element'.
>
> Now, it is this 'human element' on which the entire economics of khadi rests; and human selfishness, Adam Smith's 'pure economic motive' constitutes the 'disturbing factor' that has got to be overcome. What applies to the production of mill cloth, therefore, does not apply to khadi. Debasing of quality, adulteration, pandering to the baser tastes of humanity, are current staples in commercialised production; they have no place in khadi, nor has the principle of highest profit and lowest wages any place in khadi. On the contrary, there is no such thing as 'pure profit' in khadi.

Next year he explained at a wider level, 'What we seek to do is substitute false and non-human economics by the true and human. Not killing competition but life-giving cooperation is the law of the human being'.

Cooperation of human beings is of crucial importance and this alone can reduce the dependence on oppressive machinery.

In 1945 Gandhi wrote:

> When dependence becomes necessary in order to keep society in good order it is no longer dependence, but becomes cooperation. There is sweetness in cooperation; there is no one weak or strong among those who cooperate. Each is equal to the other.
>
> Members of a family are as much self-dependent as inter-dependent. There is no feeling of either mine or thine. They are all cooperators. So also, when we take a society, a nation or the whole of mankind as a family all men become cooperators. If we can conceive a picture of such

> cooperation, we shall find that there would be no need of support from the lifeless machine. Instead of making the greatest use of machinery we shall be able to do with the least use thereof and therein lies the real security and self-protection of society.

This spirit of cooperation he sought to realise most closely in a spinning centre:

> The centre has no dividends to make, has no exclusive care but the care of the most needy. This centre to be useful must keep itself clean in every sense of the term. The bond between it and the component parts of the vast organisation is purely spiritual or moral. A spinning centre, therefore, is a cooperative society whose members are ginners, carders, spinners, weavers and buyers – all tied together by a common bond, mutual goodwill and service.

Gandhi was not at all overawed by the success of the economic super-powers of his day. In 1926 he wrote:

> So far, it appears that the Western nations have divided all the known races outside Europe for exploitation and that there are no new worlds to discover. Among the exploited, India is the greatest victim. Japan is taking the share of the spoils no doubt. But if India and China refuse to be exploited, what will happen to the exploiters?

Next year he made an even more forthright statement: I do not believe that multiplication of wants and machinery contrived to supply them is taking the world a single step nearer its goal.

Gandhi on Open Markets

Gandhi was well aware how hidden state subsidies help the big industry and hinder the cottage industry:

> In the open market, a more organised industry will always be able to drive out a less organised one, much more so when the former is assisted by bounties and can command unlimited capital and can therefore afford to sell its manufactures at a temporary loss. Such has been the tragic fate of many enterprises in this country.

Therefore, he asked for a different type of socio-economic evaluation:

> I, therefore, maintain that, though yard per yard khadi may be dearer than mill-made cloth, in its totality and in terms of the villagers it is the most economic and practical proposition without a rival. Khadi may be interpreted to include other village industries for the purpose of a thorough examination of the proposition.

Much earlier, he had made it clear that he linked economics closely to moral and ethical principles. In 1924 he wrote:

> That economics is untrue which ignores or disregards moral values. The extension of the law of non-violence in the domain of economics means nothing less than the introduction of moral values as a factor to be considered in regulating international commerce.

In 1927 he further said: Religion to be worth anything, must be capable of being reduced to terms of economics, economics to be worth anything must also be capable of being reduced to terms of religion or spirituality. Therefore, in this scheme of religion cum economics, there is no room for exploitation.

These views were reflected increasingly in Mahatma Gandhi's perception of the role of a consumer or a buyer. For Gandhi, a buyer in need of a product should not enter the market merely to maximise his satisfaction. Instead, he should be guided to a large extent by social responsibility.

He unhesitatingly extorted people again and again to buy khadi and support it, ignoring questions such as coarseness or unevenness of the fabric.

In 1934 he wrote:

> Then take the question of standardisation. You cannot enforce it in khadi. As Rajagopalachari once remarked, a poor ordinary spinner cannot always spin thread of a uniform quality. She is not a machine. Today she may be unwell, tomorrow her child may be ill and her mind may be distracted. If you have love for the poor spinner or her child, you will not insist on having smooth, even thread always, but be satisfied with what she can give, so long as she gives her best in the condition in which she finds herself at he moment.

Earlier in 1929, he had declared: Let it be known that there are millions in India who can work at the spinning wheel for eight hours a day, and that it is impossible for them to use all the khadi woven therefrom. It is the bounden duty of good citizens of India to take off the surplus product that is turned out by these brothers and sisters of theirs.

Definition of Self-reliance

Closely linked to this is the concept of 'Swadeshi'. Literally this means 'my country' but Gandhi used the word in a much broader sense. Explaining this concept, he wrote in 1916:

> Swadeshi is that spirit in us which restricts us to the use and service of our immediate surroundings to the exclusion of the more remote. In the domain of economics, I should use only things that are produced by my immediate neighbours and serve those industries by making them efficient and complete where they might be found wanting.
>
> Interpreted in this way, Swdeshi will take India to a stage where every village of India will almost be a self-

> supporting and self-contained unit, exchanging only such necessary commodities with other villages as are not locally producable.

In 1931 he declared: A votary of Swadeshi will carefully study his environment and try to help his neighbours wherever possible by giving preference to local manufactures even if they are of an inferior grade or dearer in price than things manufactured elsewhere. He will try to remedy their defects, but will not give them up, because of their defects and take to foreign manufactures.

Gandhi explained that the concept of Swadeshi did not involve any ill-feeling towards foreigners or other related narrow feelings. He wrote in 1923:

> My definition of Swadeshi is well known. I must not serve my distant neighbour at the expense of the nearest. It is never vindictive or punitive. It is in no sense narrow, for I buy from every part of the world what is needed for my growth. I refuse to buy from anybody anything however nice or beautiful, if it interferes with my growth or injures those whom nature has made my first care. I buy useful healthy literature from every part of the world. I buy surgical instruments form England, pins and pencils form Austria, and watches from Switzerland. But I will not buy an inch of the finest cotton fabric from England or Japan, or any other part of the world, because it has injured and increasingly injures millions of inhabitants of India. I hold it to be sinful for me to refuse to buy the cloth spun and woven by the needy millions of India's paupers and to buy foreign cloth although it may be superior in quality to the Indian hand-spun.

In 1931 he warned:

> But even Swadeshi like any other good thing can be ridden to death if it is made a fetish. That is a danger that must be guarded against. To reject foreign manufactures merely

> because they are foreign and to go on wasting national time and money to promote manufactures in one's country for which it is not suited, would be criminal folly and a negation of the Swadeshi spirit. A true votary of Swadeshi will never harbour ill-will towards the foreigner; he will not be moved by antagonism towards anybody on earth. Swadeshism is not a cult of hatred. It is a doctrine of selfless service that has its roots in the purest ahimsa, i.e. love.

Swadeshi should be used to support not the products of local mills but that of village industry. In 1926 Gandhi said clearly:

> The test of Swadeshi is not the universality of the use of an article which goes under the name of Swadeshi, but the universality of participation in the production or manufacture of such an article. Thus considered mill-made cloth is Swadeshi only in a restricted sense. For in its manufacture only an infinitesimal number of India's millions can take part . But in the manufacture of Khaddar, millions can take part.

He also emphasised the voluntary aspect of Swadeshi: It is believed by some that Swadeshi could be affected by an embargo on foreign imports after the attainment of Swaraj. But that Swadeshi will be no Swadeshi. It will be a virtue practised under compulsion. True Swadeshi is the invulnerable bulwark of the nation and it can only be said to be accomplished if it is practised as a national duty.

Gandhi was very clear that the progress of Swadeshi and Khadi should continue after the end of foreign rule, as these are equally relevant to post-independent India. He wrote in 1947, the year of India's independence: We were trying through khadi to place man above the machine, rather than allow the machinery driven by electricity or steam, to be the master. We were endeavouring through khadi to establish equality between man and man in place of the enormous inequality now existing between the poor and the rich, between the high and the low, between the man and the woman. We also endeavoured to make the labourer independent of the capitalist instead of the capitalist exploiting labour and assuming undue prestige. If, therefore, what we did in India during the last thirty years was not wrong, we should now carry on the programme of the spinning wheel, with all its allied activities with more understanding of all the implications and with greater vigour.

As cloth is the most important industrial product, particularly in villages, village level self-reliant production would have established a firm basis for other village-based cottage-scale industries such as leather and footwear, but this path of development was by and large neglected in post-independence India.

Job Losses

Although the government continued to give some support to khadi and village industries, an effort supported by several Gandhian organisations, nevertheless the production of most of the cloth has remained with the powerlooms and mills or powerlooms linked to mills. As a rough estimate it may be said that nearly 72 per cent of the cloth is produced in India by powerlooms and mills, 19 per cent by handlooms and 9 per cent consists of hosiery products. Handlooms manufacture only a small share of the total cloth and most of the handlooms also depend on mill-manufactured yarn. The share of hand-spun hand-woven khadi in the overall cloth production today is very very small.

While handloom weavers suffer from many-sided problems, their most frequently voiced grievance is that of the non-availability of yarn at a fair price. The reason is that most handlooms get their yarn not from numerous charkha spinners, as Gandhi desired, but from mills. Hence, the handloom as it exists today, remains dependent on mills to a large extent. In order to increase their profits, mills frequently re-reel hank yarn (meant for handlooms and free of excise duty) into cones and divert it to powerlooms.

Due to shortage of yarn and other factors, mass distress among weavers has been reported from several places. Even in such situations, the government, in its all-out effort to increase exports, has made the situation worse for weavers by resorting to large-scale export of cotton yarn.

An expert on the textiles industry, LC Jain in 1983 estimated that during the previous decade, 5.5 million handloom workers were rendered unemployed. Similar massive loss of employment was seen in other areas of the textile industry such as the hand-printing industry.

Thus, ignoring the potential for village-based cottage-scale production has brought unemployment, under-employment and poverty to several million families – a tragedy that could have been avoided by making adequate use of existing skills of weaving, spinning and related works in villages.

As recently as 1993, Nandini Joshi, who has a doctorate in Economics from Harvard, wrote a book in Gujarati (which has also been translated in Hindi) titled, *Our Distress and Alternatives,* which argued that khadi and the spinning wheel are still practical and economically viable if only we give them a fair chance. An additional argument she advances is that khadi can help us to recover several hundred thousand hectares of land as badly needed fertile land for growing food. The mills require long and medium staple cotton which need more fertile land, irrigation and chemicals. On the other hand, short-staple cotton needed for the charkha can be obtained on less fertile land some of which is not under cultivation at present and there is no need for agri-chemicals.

Sustainable Development Goals
Seen through Gandhi's Glasses

THE FORMULATION AND prioritisation of Sustainable Development Goals (SDGs) has been a significant step in the direction of bringing out welfare goals of humanity in the form of more specific, time-bound and quantitative targets. This sets clearly identifiable goals in relation to which the progress of various countries towards better welfare, sustainability and environmental protection can be evaluated. A significant part of the development discourse is now in terms of SDGs.

It is well-known that progress in one area is related to another area. The inter-dependence of various SDGs is also clear. However, there needs to be more concern about wider trends within which the SDGs will have to be realised within the time-frame set for them. This also happens to be a time when a survival crisis is getting accentuated in terms of the very life-nurturing conditions of our planet being threatened very seriously.

This survival crisis has at least two components. One is in the form of several environmental threats led by climate change which have acquired the dimensions of a survival crisis. Secondly, there is the risk from the accumulation of various weapons of mass destruction (including the Artificial Iintelligence – AI or robot weapons, still in a relatively early stage).

This threat becomes even more ominous, given the fact that the possibility of actual use of weapons of mass destruction have increased in recent times due to a complex of factors.

If the situation in some important contexts is so ominous as to amount to a survival crisis, can we really obtain the most ambitious development goals and welfare objectives in the middle of such a situation? The SDGs have been formulated in terms which, if achieved, will amount to the most significant development achievement ever in human history at the world level. Can such a significant achievement be made in sustainable terms at a time when dark clouds of a survival crisis are gathering and growing.

This is an important question which should not be pushed aside but should be faced.

There should be no attempt to present exaggerated pictures of achievements, a trend which has already caught up in several places, with higher than actual achievements reported and accepted readily from bottom to top. There should also be no attempt to hide the fact that even real achievements can be threatened in the near future by survival crisis-related factors.

A realistic study and appreciation of the emerging situation would tell us that the prospects of steady, sustainable, durable, real progress in human welfare (and its various indicators) do not appear to be bright just now. The prospects of protection and welfare of other forms of life (other than humanity) appear to be even more dim and dismal.

In this situation, to improve the prospects of SDGs being realised, we need a much wider people's campaign for peace, justice, equality and environment protection. It is only when these value-systems become much stronger in our society and the linkages between these and SDGs are well appreciated that a firm base for real progress can be established. To achieve this, our troubled world needs to learn from various inspiring persons whose work and ideas have contributed to strengthening much to these values.

In this context, the ideas and work of Mahatma Gandhi, as well as the various social movements who have continued to walk on this path with sincerity and honesty, are of particular importance. SDGs, which involve meeting the basic needs of all people on a sustainable basis, cannot be achieved in most societies without achieving more equality and justice in terms of access to basic resources. The struggles for justice and equality must be peaceful and transparent and must involve community mobilisation to achieve stable and wider results. Various struggles coordinated and inspired by Gandhiji provide the biggest example of this in recent history. These struggles for justice were accompanied by various constructive activities, many of which are very relevant in the context of SDGs as well.

For example, the constructive programme for hygiene involved a lot of people. The constructive programme for checking alcohol consumption led to the emergence of successful grassroots action on this issue in many places. The efforts for social equality touched the conscience of the society. In the Gandhian constructive programmes aspects of sustainable livelihoods, based on local resources were emphasised. The ideal of largely self-reliant rural communities which Gandhiji emphasised (his concepts of gram swaraj and khadi are linked to this) can ensure that many SDG goals can be met in self-reliant ways with the minimum environmental burden. Community action is integral to such a way of working and can help greatly in realisation of SDGs.

SDGs cannot be realised and the survival crisis cannot be averted in a violent world. Meeting these challenges requires assured conditions of peace where good governance with community support is possible. Gandhian thinking, which emphasises peace and non-violence in all aspects of life and in resolving all conflicts, is very relevant in this context.

Important SDGs objectives in the context of environment protection cannot be achieved if consumerism trends continue to prevail. The balance of environment objectives with 'fulfilling basic needs of all' objectives cannot be achieved unless needs gets priority over greed.

Gandhian thinking with its emphasis on voluntary acceptance of frugality and simplicity as a way of life and giving creative outlets to this frugality to make it interesting, is very relevant for our times. If a frugal way of life has voluntary and willing acceptance of many more people, then this can make an important contribution to environment protection objectives. Gandhian thinking can make an important contribution to this. Examples of Gandhian community life and 'ashram' life are relevant in this context.

Saying Yes to Need, No to Greed

The concept of 'basic needs' is one of the most widely used concepts and yet this is also one of the most elusive concepts.

Everyone seems to agree that 'basic needs for all' is a highly desirable objective, but at a practical level there are several difficulties as people have very wide-ranging views on what constitutes basic needs. Finally, there is the even more difficult question of what should be the aim of society once 'basic needs for all' have been realised, if and when that happens.

In a poor community, however, there is likely to be less controversy on defining basic needs. This community is likely to define 'basic needs' of a household somewhat along the following lines:

1. Certain minimum norms of food (including clean drinking water) which are necessary to keep the body in a healthy state.
2. The least expensive housing, sanitation, furnishing and clothing that will provide adequate protection from weather extremes and is otherwise essential for protecting the family's health and
3. Access to some basic health services and school-level education.

This appears to be an eminently reasonable definition of basic needs, something which appears achievable for all people. However, a large number of people do not agree with this definition of basic needs. Recently the *Daily Telegraph* asked a family of six having six computers, three cars, nine TVs (including one in the kitchen), piped music throughout the house and almost every domestic appliance to do without these gadgets for just three days. By and large, the family found it extremely difficult to cope and waited eagerly for the three days to be over. One member was so appalled by the prospects of living without all these gadgets for three days that she refused to participate in the experiment.

Many such examples can be given to show that the concept of basic needs has a very bloated and exaggerated meaning for an increasingly large number of people.

Goods and services without which they cannot live (or think they cannot live) are so many that their 'needs' include what millions of people would regard as the biggest luxuries. Why does the modern man's perception of a satisfactory life include such a wide range of goods and services, a range that appears to be increasing all the time? Is this because we've lost something precious and try to fill this gap in our life by increasing acquisition of goods and gadgets? Is this a desperate, but also futile, effort to fill a social and spiritual void with material goods?

Generally, it can be said that the more the breakdown of human relationships, the grater the dependence on purchased goods and services. In a joint family, so many additional needs are met by serving each other all of which have to be paid for when the joint family breaks down. This is even more true at the wider community level. The more inward looking and isolated we become, the greater our dependence on material acquisition.

Our life has become extremely busy and we are always in a rush. We do not have enough time for human relationships. So, we try to replace this by buying expensive things for our near and dear ones, without even thinking whether the replacement of genuine care by material goods is acceptable to them.

It is amazing how everyone appears to be in such a big rush, and how little actually gets done to improve life. Several of us manage to convince ourselves that our very busy work schedule is of great significance, and to support this, we hasten to acquire as many goods and services as possible which we perceive as necessary for this life-style. A similar bloated view of the importance of our 'exhausting' work schedule convinces us that many entertainments are equally essential to recover from this entirely self-inflicted exhaustion.

What we avoid generally is a truly honest analysis of how useful this extremely busy work schedule is really for the creation of a better world? A very frank and honest assessment is likely to bring disturbing conclusions to most of us and force us to reconsider our life-style.

Taking Advantage of Human Weakness

A very strong force that leads to a bloated sense of needs is the constant advertising that bombards us from all side at all times. Advertising does have a role as far as giving relevant information to people about useful products is concerned, but *present day advertising goes much beyond this limited role and constantly makes us yearn and crave for goods we do not really need.* Advertisers take advantage of every conceivable human weakness and play on human emotions in the most reckless way to sell all sorts of products of dubious merit, or even clearly harmful products. Billions of dollars are spent every year in the advertising of liquor, cigarettes, junk food and useless or even hazardous drugs, cosmetics and pesticides.

One particularly perverse aspect of advertising is how it distorts serious problems in such a way that solutions in the form of genuine social change are replaced by commodities that can be sold.

Partly as a result of our own weaknesses and partly as a result of heavy advertising, we've been steadily commoditising our life, vainly trying to fill social and spiritual voids with material goods. Understanding this is the first step towards changing our life-style in such a way that our needs are not in contradiction with the social objectives of environmental protection and meeting the basic needs of ALL people.

Suitable changes in government policies and public utilities can also contribute considerably to reducing excessive private spending. Availability of good public transport can reduce the need for privately-owned vehicles. Community gatherings and celebrations can reduce the need for privately organised expensive entertainments.

Thus, a combination of changes in personal life-style, enhanced role of community life and relevant changes in government policy can help us to modify our understanding of basic needs in tune with desirable social norms.

While the well-to-do household's perception of basic needs will still be different from a poor family's perception, it will not be too different. Inequalities will still exist, but not too much. The present-day society and the state of environment simply cannot afford a very bloated view of needs. It will be difficult to fulfil an exaggerated sense of needs for everyone without inflicting heavy damage on already badly mauled environment. So reaching a realistic understanding of needs is essential if we want to create a world where basic needs of ALL can be met on a sustained basis without damaging the environment too much and protecting the interests of future generations as well.

The idea that the production of material goods and services is desirable only up to a point, beyond which the pursuit of this objective should be given up voluntarily may appear strange to some growth addicts but it is deeply embedded in teaching of several spiritual leaders and social reformers who have enjoyed very large followings. If we can get rid of the dominance of some narrow-minded economic thinking of relatively recent origin and give due importance to the great spiritual and social movements of human history, then there are many sources of support for this idea. By drawing on these inherent strengths of various cultures and societies, we can build a new world which is free of greed and hence of most type of violence, where the basic needs of all people are met, where human relationships are valued, where there is genuine regard for other forms of life and for protecting natural environment for this and future generations.

Simple Life and Equality

Two Pillars of Environment Protection

COMMITMENT TO THE protection of environment demands a change of approach to life from acquisitiveness to simplicity. But to call upon the people to make this change would be meaningless unless it is accompanied by a call for another commitment – the commitment to equality in socio-economic life.

For the message of a non-acquisitive and simple life-style to have any meaning, there should first be a commitment to change the injustice and inequalities of this world – whether these are between person to person, or nation to nation. Any talk of changing the acquisitive lifestyle without attacking the injustices and inequalities of our world is meaningless.

However, the reverse of this is also true.

A society which has established economic equality to a

considerable extent but is not alert to the need of spreading the message of a simple non-acquisitive lifestyle is asking for trouble of more than one kind.

It is simply not possible to provide a consumerist and acquisitive life-pattern (following the norms set by the elites of developed countries) to all people. If such are the ambitions to which people have been led, then at some place the norm of equality will break. A section of the people will be excluded from this dream world or the avaricious nation will try to exploit other countries to satisfy the greed of its own citizens. This is an even worse, an even more unjust, violation of the equality norm. If the equality norm is not broken, then the desperate nation trying to tap its natural resources too rapidly and on too large a scale will end up creating ecological havoc which in turn will wipe up whatever gains that have been obtained in the standard of living.

Very lavish and comfortable life-style can exist either for a minority in highly unequal societies or else these have to be supported by the exploitation of the natural resources and workers of other countries. It would be very unwise to promise the sky to workers while mobilising them for fight against an unequal and unjust system. Instead, the concept of a satisfactory but simple life-style should be stressed from the outset. Without this additional concept of a non-acquisitive lifestyle, an equal society cannot be sustained. It will either not remain true to its own objectives, or build up ecological catastrophes, or collapse under the weight of its own contradictions, as has been tragically seen in some countries in the recent past.

On the other hand, to preach for a non-acquisitive and simple lifestyle without attacking inequalities and injustices is sheer hypocrisy. This is empty talk and will not take us anywhere.

Equality and simplicity are two important concepts which acquire their real significance only when taken together with each other. One without the other will not take us far, but the two taken together will help greatly in building a new, a better world.

Thus, equality and simplicity (self-restraint) are the two pillars on which the foundation of sustainable development can be built. To be able to achieve this, very powerful vested interests have to be confronted and overcome, and very deep-rooted economic and political changes both at global level and at the local level have to take place.

A well-thought out, dedicated and sustained campaign to convince more and more people that both happiness and security are best achieved by pursuing sustainable development based on equality and simplicity which are the biggest need of our times. Once most people are convinced of this, even the most powerful vested interests will not be able to check change towards the path of sustainable development.

The concept of development needs to be freed from the terrible misconception of more and more accumulation and consumption. As long as production of material goods is increasing in a society to meet the basic needs of all, this is justified and in conformity with the welfare of this society. But significant production and consumption of material goods beyond this cannot be justified and should be called 'wasteful destruction', not development.

Development should be redefined as the collective endeavour to reduce distress and suffering of all forms of life as much as possible. Some distress and suffering will always be there, but most of the distress and suffering of this world is caused by avoidable causes. *Development needs to be seen as the efforts of human beings to check these avoidable causes of distress and thereby reduce the distress of all forms of life as much as possible.*

This endeavour should take good care of the needs of future generations. We should protect the resources base so that future generations of human beings and other forms of life can meet their basic needs equally well or in even better ways. *This is the essence of sustainable development.*

Need for Land Reforms
Importance and Growing Official Neglect

Solutions for Land Inequality

IF THE MOST important factor in the context of poverty, hunger and the related discontent in India has to be identified in just one word, then this word undoubtedly will be 'land'. It is injustices and distortions relating to land distribution which are the single most important cause of poverty, hunger and rural unrest not only in India but in many other countries as well.

The solution clearly is land reforms, which may be defined here as ensuring of proper land rights for all rural workers and peasants to such an extent that their sustainable livelihood is well protected. This may involve distribution of farmland, distribution of other cultivable land, protection of existing land rights of peasants and other measures.

Land scarcity is acute in India. India has 2.4 per cent of the world's geographical area and 16 per cent of the world's population. It has 0.5 per cent of the world's grazing area but has over 18 per cent of the world's cattle population. India has nearly 18 million landless farm/rural labour households, a total population of about 100 million.

In these conditions, providing a secure land base to landless and near-landless peasants is of the greatest significance to provide them food security and to reduce/remove their poverty.

Land reforms are an essential and an extremely important component of any paradigm of development that sincerely wants to remove poverty and provide food security in conditions prevailing in India.

After independence, India's land-reform efforts started with the aim of helping landless farm toilers to become small peasants owning small plots of farmland. Seventy years later, we can see that this effort did not succeed – the achievements are so small as to be negligible. On the reverse side, there is a much more massive drive to turn small and middle peasants into landless workers. This takes the form of massive displacement, land-grab, indebtedness, high costs, ecological ruin and other forms of deprivations.

This is perhaps the biggest tragedy of India's agricultural experience that the noble aim of helping the landless farm workers to become farmers got reversed into the harsh reality of turning farmers into landless workers.

Therefore, any comprehensive land-reforms effort should include both these essential components – providing land to landless peasants and protecting land rights of existing farmers.

Such a land reform effort can be more useful than any other programme in reducing poverty, increasing productivity, ensuring food security as well as bringing peace and justice to Indian villages. What is more, such a programme can create the most enthusiastic mass base for a programme of ecological regeneration.

The Planning Commission, India's highest planning authority (now abolished), had expressed its agreement with this key role of land reforms. In the Tenth Plan document the commission had stated, 'In an economy where over 60 per cent of the population is dependent on agriculture, the structure of land ownership is central to the well-being of the people'.

While acknowledging the crucial importance of land reforms, this official document also admitted the government's growing reluctance (even opposition) to take up this important task. 'Land reforms seem to have been relegated to the background in the mid 1990s. More recently, initiatives of State governments have related to liberalising of land laws in order to promote large-scale corporate farming.'

The report of the Working Group on Land Relations for Formulation of the Eleventh Five Year Plan (the WGLREP Report) stated even more frankly, 'There are serious concerns that the phase of economic liberalisation may be accompanied by further erosion of land-reforms'.

These concerns are summarised by Pushpendra:

> The pro-liberalisation lobby is for the removal of all restrictions on leasing, upward revision of ceilings on holdings, concessions to the industry for investing in rural infrastructure, unrestricted entry in rural areas for national and multinational companies, agro-farm houses, food and fruit processing industries to lease-in or purchase land through direct interaction with the peasants, speeding-up the acquisition process by the government, export-led and demand-driven agriculture, and for the removal

> of all restrictions on conversion of land from farm to non-farm activities. There is also immense pressure on the government to make available land to private parties either through land acquisition process or from its own pool of land meant for distribution to the landless and the small and marginal farmers. Land being a State subject, some State governments have already made changes in this direction in their legislations. This has given birth to several apprehensions in certain quarters.

In this dismal situation, the many-sided benefits and significance of land-reforms need to be re-emphasised.

Sustainable Productivity, Poverty & Environment

1. Sustainable Production

Small farmers are likely to cultivate their fields more intensively and carefully, working hard to improve land fertility and productivity. Particularly in the context of sustainable farming practices and organic farming practices, it is the ability of farmers to carefully nurture and cultivate their fields, prepare organic manure and pest-repellents which is more important. This is clearly better achieved on small farms and family farms.

Land reforms are also helpful for protecting land fertility, sustainable use of land as well as wastelands development as hard-working poor peasants (including first generation farmers) are likely to work more sincerely for soil and water conservation and related works.

For somewhat similar reasons, land reforms can also lead to a rise in farm productivity, that too on the farms of the poor. This view is supported by several international studies.

In a widely-quoted publication titled, 'Agriculture Towards 2000', the Food and Agriculture Organisation (FAO) has emphasised that more equal land distribution is likely to increase productivity of land, 'It is important to stress here that yields per hectare are as high on small as on large farms or, under traditional agriculture, even higher. With a few notable exceptions, total output per hectare is higher on small farms, chiefly because their intensity of land use is higher'.

This view of FAO is supported by a six-country study by the International Labour Organization (ILO) which estimated that:

> If land were equally distributed among all agricultural families (including the landless), and the new equal holdings achieved yields equal to present holdings of the same size and used a similar level of inputs, food output could potentially rise by anything from 10% (Pakistan) and 28% (Colombia and a rice-growing Malaysian region) to 80% in northeastern Brazil. Such a radical redistribution is, of course, rarely attempted – but the figures indicate the theoretical potential.

This view that greater equality in farmland distribution will increase productivity is also supported by a World Bank publication titled, 'The Assault on World Poverty, Problems of Rural Development, Education and Health'.

This study notes that in Thailand, plots of two to four acres produce almost 60 per cent more rice per acre than farms of 140 acres of more. An analysis of the differences in the value of output on large and small farms in Argentina, Brazil, Chile, Colombia, Ecuador and Guatemala revealed that small farms were three to fourteen times more productive per acre than the large farms.

France's Lappe and Joseph Collins examined the data of net income per acre by farm size in the US from 1960 to 1973. In 12 out of 14 years the net income per acre was greater on the (smaller) family farms.

However, to ensure that the potential of higher productivity is realised, small farmers and particularly new land allottees should get the necessary help for irrigation, water and soil conservation, composting etc.

2. Fighting Poverty and Hunger

The landless people are the poorest people in India's villages and if they get land to grow food – then clearly, this is the most important and durable contribution to reducing poverty and hunger.

The fact that per acre productivity can be higher on small farms also increases the role of land reforms in reducing poverty and hunger and contributing to food security.

The above-mentioned FAO report said in the specific context of India: Redistribution of only 5% of farmland in India, coupled with improved access to water, could reduce rural poverty levels by 30% under what they would be, so that in Indian conditions land and water reform would be a key approach.

3. Peace and Democracy

Land inequality, unjust deprivation and land grabbing have been repeatedly found to be one of the biggest causes of distress and discontent in rural areas. This is the biggest factor behind the upsurge of violence in many areas. Well-enacted land reforms can contribute greatly to reducing this discontent and the violence arising from this discontent.

In the interests of peace and democracy, it is of the greatest significance for government to give adequate attention to land reforms. The government should respond adequately to just demands made by peaceful movements. Otherwise widespread discontent leading to violence is most likely.

4. Protection of Environment

As already stated, small farms and well-motivated small farmers provide more conducive conditions for careful cultivation and soil-nurturing practices needed for sustainable, organic farming. Millions of beneficiaries of land reforms in particular (but not excluding others) can be mobilised to repair traditional irrigation sources, take up various soil and water conservation works, protect and regenerate forests and take up new afforestation work. A massive environment protection and regeneration work with the enthusiastic support of land reform beneficiaries will achieve more for sustainable livelihoods and food security in India (as well as many similarly-placed developing countries) than anything else.

5. Dignity

Time and again it has been observed that land-entitlements contribute to a sense of dignity among people, particularly the poorest people in villages who have been denied land for so long. In this context, the entitlement of homestead land is very important and also some cultivation land in addition to homestead land.

6. Strengthening Rural Community

Rural community has a stronger and more harmonious base where farmers have more or less equal land holdings or at least where land inequality is not high. On the contrary, community life will have a weaker base and more room for conflict if inequalities are very high. Hence land reforms contribute to a better community life.

The Perspective of Land Reforms

In this entire discussion, the perspective of land reforms needs to be brought in. The government has repeatedly gone back on its promises of large-scale, significant redistribution of farmland among the poorest landless (or near landless) people of rural areas. At least some land was distributed and there was hope for more in the early years of a free India.

The massive assault on farmland and displacement of people, particularly farmers, in the recent decdes has a very adverse impact on the possibilities of land reforms as well. When even land-owning farmers are losing their land on a vast scale, who will care to give land to landless peasants?

When land is acquired for various projects, landless farm workers, tenants, sharecroppers, artisans etc. are often left in the lurch as they lose their livelihoods but don't have any land ownership papers to claim compensation and other 'benefits'. Although more recently, some efforts have been made to partly alleviate some of these problems, the fact remains that on the whole assaults on land, commons and forests also harm the poorest landless people in various ways while also reducing drastically the chances of any land distribution among them. Landless or near landless farmers who have got their land after years of hard efforts for land reforms and social justice are also known to have lost their fields in the various assaults on farmland. Many of them have worked very hard to reclaim the poor-quality land distributed to them and make it fit for cultivation.

So, the perspective of the landless and near landless people and of the continuous need for land reforms should be brought in adequately into this struggle against the assault on land and forests. And it is clear that from this perspective also, the assault on farmland should be opposed and resisted.

Those movements which have been struggling only on behalf of the landless and the near-landless should now widen their horizons as with the assault on farmland many farmers also face the threat of becoming landless.

Proposal for a Minimum Holding

It is in this wider context that the proposal for a law ensuring minimum land holding for all long-term resident rural households acquires many-sided significance. Such a law could lay down, for instance, that all long-term residents of a village will have rights to homestead land and at least about an acre of cultivation land. However, the size of the land holding can differ in hill and plain areas, and in irrigated and desert areas.

Such a law is justified at any time on the basis of social justice, but its importance has increased at a time of the widespread assault on farmland as this can give some minimum security not only to landless people but also to others now threatened with landlessness. Even when such an assault cannot be stopped, this law can compel the authorities to ensure that at least some minimum land base of any family is not disturbed (or has to be provided at another nearby place).

The assurance of a minimum land base for all rural households will provide multiple benefits:

1. This will help significantly in poverty alleviation and food security in the context of the poorest families and thereby help in achieving top priority national goals.
2. This will help poorest families to explore additional livelihood opportunities including food crop production, animal husbandry and cottage scale crafts.
3. This will improve health and hygiene of poorest families

as they have space for toilets, bathrooms and a separate place for dairy animals or other animals.

4. This will strengthen rural communities as without a land-base, the poorest families can be easily lost to the rural communities in these times of greater dependance on migrant labour.
5. Several rural households have said that even a small plot of land gives them a certain social dignity.

Hence the proposal for a minimum land holding deserves immediate attention as a means of promoting food security, poverty alleviation, social security, health, hygiene and community ties. *Strengthening the viability of the farm, the farmer and the rural communities is a very Gandhian concept.*

The Good Earth
Farmland, Commons and Forests

INDIA HAS ABOUT 16 per cent of the world's population and 2 per cent of the world's land, as I have noted earlier. The livelihoods of nearly two thirds of India's population are linked to farmlands, village commons and forests.

Given the scarcity of land as well as the extent of livelihoods linked to it, it is of the greatest importance to plan land-use in such a way that the base of sustainable livelihoods is protected. The importance of this task has further increased immensely in recent times of escalating threat of climate change as sustainable livelihoods, based on greenery, organic farming and more, are much more beneficial for reducing greenhouse gas (GHG) emissions compared to a situation where these livelihoods (and the base supporting them) are uprooted.

However, precisely at the time when the importance of protecting 'green livelihoods' and their base (particularly farmland and forests) is the greatest, most tragically the assault on agricultural land, commons and forests is also the greatest. As the path of capitalist development spawns increasing inequalities within the country, and also gets more forcefully linked to the forces of globalisation, the nature of such growth and linkages necessarily create conditions of a relentlessly growing assault on green, sustainable livelihoods and their base.

To make this clearer, we may classify the various factors due to which the base of sustainable, land-based livelihoods (mainly organic agriculture, horticulture, animal husbandry etc. and mixed farming systems based on a combination of these) can be threatened.

1. Even in a reasonably balanced path of development, with the passage of time with a growing demand for industrial goods, supporting infra-structure and some urbanisation, there will be a pressure on some agricultural and forest land. However, this can be managed without very serious damage – particularly when there is a well-thought policy for saving fertile farmland and natural forests, and when this policy is implemented in participation with rural people.
2. The situation gets more difficult if the path involves extreme inequalities, as in this case there is excessive pressure on production of luxury goods, luxury housing, hotels, resorts etc.
3. If the development path gives less priority to farmers, shifts terms of trade against agriculture, and provides them little insurance in times of greater uncertainty, then it become easy for land speculators to uproot farmers.

4. If land speculators, mafias, dons are allowed to flourish and act in collusion with corrupt officials and politicians, then the prospects for excessive land grabbing in the guise of 'development' work increase greatly.
5. When the national economy is more and more deeply drawn into the web of globalisation, then the pressure to meet the demands of global elite greatly increases the pressure on sustainable livelihoods and their base.

The global demands of minerals can be particularly huge, but there are also pressures for various industrial or other goods commanded by globalization-based specialisation for various countries/regions. This often pushes more polluting, 'dirty', displacement-causing activities to poorer countries, bringing super profits to a few elites but misery to most people.

Land Acquisition

While any threats to sustainable livelihoods and their base arising from factor (1) can be managed within reasonable limits, factors (2) to (5) – particularly when combined together – can increase these threats manifold and these threats can even become limitless, given the pull of speculative forces and global demand. In such a situation, any planning to protect the base of sustainable livelihoods becomes very difficult because firstly, the global pressures cannot be predicted and secondly, powerful interests of speculators and land-grabbers create serious problems for any such planning.

Therefore, any serious effort to challenge the threat to sustainable livelihoods based on farmland, commons and forests should also resist factors (2) to (5) because if these forces are not weakened, then the assault on sustainable livelihoods and their base will continue to grow. However, government policy in India at present is not based on any such wider commitment.

Within the framework of the democratic system in India, the government's main concern is not to check the assault, or the extent of the assault, on farmland and forests but rather to manage and contain the distress and discontent this is likely to cause among farmers, pastorals, forest-produce gatherers and other sustainable livelihood occupations.

In States like West Bengal, governments have seen how volatile issues like land acquisition proved and led to serious setbacks and electoral defeat for well-entrenched ruling coalitions. Therefore, the entire effort is to think up a package of compensations and incentives which will satisfy the farmers who face the threat of land acquisition. Towards this end, the State governments of Haryana, Gujarat and Uttar Pradesh have come up with their own versions of 'ideal package' and other States may soon follow.

Now attention is more focussed on the national-level efforts to provide a 'model' law on land-acquisition as well as rehabilitation of displaced people. While various provisions of these laws are hotly debated, without challenging factors (2) to (5) mentioned above, the impact of any legislative measures on reducing the threat to sustainable livelihoods can only be limited.

Here it is important to bring out the basic difference between two kinds of objectives:

> Objective (a) Somehow reducing the discontent and protests against land acquisition without necessarily reducing the extent of land acquisition.
>
> Objective (b) Focusing on the need to reduce significantly the assault on farmland, commons and forests and the sustainable livelihoods based on these.

These are two very different objectives and should not be confused with each other.

The government appears to be mainly concerned with the first objective, and many senior persons in the government may not even accept the second objective as desirable at all. In fact, a very senior minister said some time ago that the ideal development of the country he has in mind is for 85 per cent of the people to live in cities (compared to the 30% urban population today).

However, for several social and environmental movements, objective (b) is more important. Both from the point to view of protection of sustainable livelihoods and environment protection (as well as overall welfare). Clearly objective (b) is more important and more basic.

1. If fertile farmland, commons and forests are protected – then sustainable livelihoods for many, many generations will be assured.
2. In a world where food security is increasingly threatened, this will contribute to adequate availability of food crops, milk and several kinds of nutritious food available in forests.
3. At a time when biodiversity is badly threatened, this will contribute to the protection of bio-diversity.
4. This will contribute to the strengthening of rural communities which are threatened in many ways but make available a better, more sustainable pattern of life, if given a chance.
5. This will help to reduce GHG emissions. The pattern of sustainable development based on organic farming and green cover is always better than the pattern of mostly polluting industrial and infra-structure development

which emerges after uprooting the green cover. This aspect is likely to become more and more important in times when climate change is becoming an increasingly crucial issue, indeed a survival issue.

6. If the assault on farmland is stopped, then the possibilities of land-reforms or distribution of farmland among the landless rural poor can also increase. There is wide agreement that for reducing rural poverty, land distribution among the poorest landless people can prove one of the most effective and stable means, while other options may offer only a temporary relief.
7. Also, it should be emphasised that large-scale displacement from land is likely to lead to large-scale unemployment.

Capital-intensive industrialisation can provide very limited employment to those uprooted from land. Unlike in the case of Europe's much earlier industrialisation, the prospects of uprooted people being able to migrate to other countries or continents don't exist now.

Hence what needs to be emphasised is the great necessity of protecting sustainable livelihoods and its base – fertile farmland, commons and forests (and of course the water-sources which support this greenery).

Short-term remedial measures or even appeasement of affected farmers by hiking cash compensation at selected places (or by other such means) is not the answer. The focus should be on challenging and checking the paradigm of development (or is it destruction?) based on a massive assault on the base of sustainable development.

How to achieve this in the face of the ganging up of powerful interests to support this assault, and to try to justify this as an inevitable or essential part of 'development'?

Some suggestions for strengthening these struggles and the overall effort for checking the assault on farmland, commons and forests.

a. Bring the various struggles to protect the 'Good Earth' closer to each other, with a special emphasis on various peaceful struggles, so that they can contribute to each other's success.

b. Mobilising pro-people professional expertise so that well-researched, comprehensive, cost benefit studies for these various projects can be made which bring in adequately the aspects of food security, sustainable livelihoods of rural communities, GHG emissions, bio-diversity etc.

c. Setting up State-level and national platforms for these struggles to take up issues of common interests and to provide them support.

d. Linking these national initiatives to international organisations working on sustainable development alternatives, promoting food security and checking climate change.

e. There should be a well-organised campaign for *alternative paradigms of development in which industrialisation doesn't uproot farmers.* Instead, farmer households continue to retain small plots of cultivation land while some of their members also participate in non-displacing, non-polluting, labour-intensive small and medium sized industrial units with a special emphasis on cottage-scale rural units.

If local to State to national to world-level linkages can be made and if the perspectives of food security, poverty alleviation, environment protection, sustainable development and checking climate change can be combined in an effective way, then the prospects of checking the assault on poverty can improve significantly. This entire effort should, of course, include well-organised media campaigns, establishing regular dialogue with legislators and political parties, academics and opinion-makers, as well as public interest litigations.

Gandhian Solutions More Relevant

Key to Reducing Rural Distress

IT IS GOOD that there has been a lot of discussion around farmers' problems in recent months. Some very useful suggestions have emerged at the level of the official policy. However, two issues appear to be dominating the discussion.

Firstly, the much-discussed loan-waivers.

Now there is no doubt at all that farmers need relief from debt, not just from bank-loans but also from the loan of private moneylenders. But alongside, we need to raise the highly relevant question – what is the system that pushes the farmer into debt and why the system has aggravated to increase the indebtedness of farmers in more recent times (despite selective loan-waivers being announced from time to time)?

Are we also doing something significant to break the system which inevitably increases indebtedness? If this is not done, then despite loan-waivers – even bigger ones – the problem of indebtedness relating distress will again acquire serious dimensions after some time.

The second policy option, which dominates discussion, is regarding increase of support price. Again, there is no doubt that this is needed, but again there is a larger context and a larger question. The crucial question is – what kind of farming system will the increase in price support?

If it helps farmers who are involved in ecologically friendly and sustainable farming practices, then this increase is most welcome. On the other hand, if such an increase perpetuates a basically unsustainable and ecologically harmful system by providing an artificial support, then the wisdom of such a price-increase is in doubt.

For example, growing excess of sugarcane in a water-scarce area (and also having liquor factories linked to large-scale sugarcane cultivation) is clearly unsustainable. It is basically harmful for the environment and people of the region and so the system is bound to collapse. But politically powerful lobbies can exert pressure to increase the price of this monoculture crop even in such a situation and so the unsustainable, harmful cropping and industrial growth linked to it is perpetuated by using public funds wrongly.

This is just one example. The broader question is, whether public funds should be poured into providing an artificial long-life to unsustainable, expensive, ecologically disruptive systems or should the same funds be used to support a change towards sustainable, low-cost, eco-friendly, self-reliant systems?

There are very powerful lobbies which not only want the existing unsustainable, ecologically disruptive and harmful practices to continue, but in addition they are pushing for even more disruptive practices that benefit them or increase their dominance. Should the government make their task easier by increasing the support-price for the crops or crop varieties or crop rotations grown under these ecologically disruptive systems?

The key issue is that we need farming systems which are sustainable, extremely low-cost and also self-reliant (to the extent possible).

One main emphasis is to find decentralised and local-specific solutions, keeping in mind local land, soil and water conditions, while keeping in mind the larger priorities of agro-ecology, food security and sovereignty.

Farming science should mean never going against nature, but instead observing natural processes carefully to see how these can be carefully harnessed to help the cause of farmers and farming without harming other living beings and environment. A related principle is that no poison should ever be introduced and released in the food-system and in the food chain.

For these good ideas to work in the interests of the good earth, the government should have very pro-village policies and spend liberally but carefully on water conservation and increasing the green cover.

A question which many farmers' organisations ask is whether such a system can also bring enough cash to farmers, as beyond providing healthy food for themselves and others they have cash needs. The answer for this often-asked question is three-fold. Firstly, it is not a matter of gross cash receipts but of net income. If good farming is so low-cost as to avoid debts, then the net income is likely to be better. This net income will increase as the positive results of protected ecosystems emerge in the course of years.

Secondly, there is enough scope for generating a wide range of other eco-friendly jobs at the level of rural areas and para-urban areas which can be more accessible to all rural families and help to increase cash income. Such wider changes in economy for decentralised employment increasing initiatives are also needed for regeneration of villages.

Thirdly, social reform efforts are badly needed to curb consunption of liquor and all other substance abuse, to curb dowry system and all wasteful spending on certain events and also to restrict wasteful consumerism .Without such social reform initiatives, it will be difficult to reduce indebtedness and economic problems.

The problem of rural distress is multi-dimensional, and the solution also has to be multi-dimensional. We should try to reach a broad agreement at least on some basic issues so that some important policies and initiatives can be taken up without further delay.

As eco-friendly farming is an important part of the solution, we should try to also get the funds available for reducing greenhouse gas emissions for such programmes.

The government must also give highest priority to rural areas. Hence it should be a possible to mobilise higher resources for rural areas, but it is even more important to utilise these funds carefully. So far, funds have been largely used to a large extent to fund unsustainable, ecologically disruptive, wasteful practices. We need to use our limited available resources for sustainable, eco-friendly practices which are also very low-cost and most suitable for small farmers. The agenda of land reforms should be revived so that landless households can also get at least some land and we can also have them as small farmers.

Small Farmer-based Agriculture: Self-reliant, Sustainable

While reducing farming costs is of the greatest importance for the success and sustainability of small and medium farmers who constitute over 90 per cent of the farmers in India, it is surprising that this has been neglected at the policy level to an alarming extent.

In fact, even in more recent times when farming issues have been very extensively debated , the focus has been on remunerative prices and loan waivers and very less on the considerable possibilities of reducing costs of farming.

Despite the skills and hard work of farmers, the yields of farmers depend on several factors including weather conditions and can never be fully assured. Most of the farmers are low-resource farmers whose capacity to absorb losses is very low. Hence it is obviously very sensible to adopt those farming strategies and technologies which can help farmers to obtain reasonable yields while keeping costs at very low levels.

The most obvious way of keeping farming costs at the lowest levels is to encourage and support the self-reliance of village communities as much as possible, something which is very compatible with the Gandhian principle of self-reliant villages and rural communities.

The government can contribute to this by providing adequate resources to rural communities to take up water and moisture conservation and protection of forests and pastures as well as planting of indigenous trees in a big way. This helps to create a strong base for self-reliant and prosperous communities. Villages can be certainly self-reliant in preparing manure and compost as well using other methods of fertilising their fields which do not involve any cash spending. (In this context, government fertilizer subsidies do just the opposite, promoting use of inorganic fertilisers.) Mixed farming systems and crop rotations offer ways of providing nutrients to soil while also reducing the possibilities of the crops being damaged by pests and disease. Draught power can be obtained by balanced development of animal husbandry.

Some people, without going into detail, would say simply that organic farming by avoiding chemical fertilisers and pesticides can help to reduce costs. But at the same time, we should not forget that some aspects of organic farming can also increase costs. For example, if the spread of organic farming involves expensive certification procedures – then this also is going to create new kinds of difficulties for farmers. So, in the context of the needs of small farmers and rural communities of India – what we need is not just any organic farming but organic farming which specifically is low-cost and self reliant.

This certainly does not mean that modern technology will not be used but only that modern science and technology will be used in such ways that promote the self-reliance of rural communities and help to meet their need for those eco-friendly methods which are also extremely low-cost.

Science and technology can be used to advance the interests of big business, but in different ways science and technology can also be used to advance the interests of self-reliant communities of small farmers. Mangal turbine, which has been innovated by a farmer scientist Mangal Singh, is a good example of a modern technology that promotes the self-reliance of rural communities as it enables farmers to lift water from small rivers and streams without using diesel and electricity.

Unfortunately, during the last five decades or so, the government has not followed this approach of promoting the self-reliance of villages or very low cost methods of eco-friendly farming.

Rather, the farmer has been treated as someone from whom others should earn as much as possible. So, in the name of helping farmers and farming, those policies and technologies are promoted from which various business interests can earn a lot but which increase the costs and economic burden of farmers. Some of these technologies promoted by the government also have various adverse impacts on ecology like harming the natural fertility of soil or killing friendly insects and micro-organisms or overusing water. As these adverse impacts manifest themselves over a period of time, the problems of farmers go on increasing and to cope with these problems their costs go on increasing all the time. (For example: use of endosulfan at one time. The Kasaragod corporation began aerial spraying of the pesticide in its cashew plantations spread across 15 panchayats in the district in 1978 and its application continued till 2001.)

To somehow justify these wrong priorities and policies, a myth has been created that certain policies and technologies may be very costly in economic terms and even more so in ecological terms but these help to increase food productivity or yield per hectare and for this reason, we have to simply accept their high economic and ecological costs.

However, such an assertion is not supported by the actual evidence.

In the 12th Plan document the Planning Commission had published the annual increase of farm yield data for various crops before and after the famous Green Revolution. As is well known, before the green revolution very little chemical fertilisers and pesticides were being used while in the green revolution years a lot of chemical fertilisers and pesticides started being used and this increased very rapidly.

Also, before the Green Revolution, farmers were using

basically their own seeds of various diverse traditional varieties of crops while during the green revolution they became dependent on cash purchased seeds of new HYVs (hybrid varieties and monoculture seeds).

Data Tells the Story

The purchase of tractors also increased rapidly, generally on the basis of loans carrying high interest rates. In other words, while before the green revolution the farming was close to the concept of low-cost farming, after the green revolution the costs of farmers increased very rapidly. In this perspective, the data presented in the 12th Plan which compares the crop yield during the period 1951-52 to 1967-68 (pre- green revolution) with the period 1968-69 to 1980-81 (during green revolution) is very significant.

This data tells us that in the previous period the average annual growth rate in yields per hectare (AAGRYH) for wheat was 3.7 per cent, in the later phase this was lesser at 3.3 per cent. In the case of rice, the AAGRYH for rice was 3.2 per cent while in the later period this was much lesser at 2.7 per cent. In the case of maize the AAGRYH was a robust 4.8 per cent in the former period but in the later period, this had declined very significantly to 1.7 per cent. In the case of pulses, the AAGRYH in the previous period was placed at 2.3 per cent but in the later period this had tumbled to a miserable minus 0.2 per cent.

This is very significant data which shows that in the case of most important food crops, despite the very substantial increase in economic costs, the average annual growth in yields per hectare actually declined in the Green Revolution period compared to the pre-Green Revolution period. (For years, the

Green Revolution has been a holy cow for the bureaucracy and the scientific community in the country, but cultivating water intensive,chemical intensive rice in Punjab has had its cost, very apparent today in the State's farming crisis. Criticising the Green Revolution has been simply a big No, no.)

Now the time has come to challenge the long-perpetuated myth that technologies and policies which have high economic and ecological costs are nevertheless needed for increasing farm productivity and production.

Instead, there is enough evidence to show that technologies which harm ecology, which harm the basic soil and water resources and are poisonous for pollinators can never increase farm productivity in a longer time horizon. (Like genetically modified GM cotton promoted in Gujarat.)

Several scientists have time and again tried to demonstrate this, but their voice is less heard. Former director of Central Rice Research Institute, Cuttack, RH Richharia was widely recognised as India's most prominent rice scientist. He carried out pioneering research work which revealed that tribal farmers and traditional farmers who were considered very backward by the modern research establishment were actually getting yields equivalent to or higher than green revolution varieties from some of their traditional varieties and that too, without using chemical fertilisers and pesticides. But such research was stifled as otherwise, it would have been difficult to justify the adoption of more expensive technologies on a vast scale.

The Scary GM Revolution

Now after a gap of several years in the name of a second Green Revolution there is a another big move to spread genetically modified or GM crops which can bring huge gains to some of the biggest multinational corporations but which will further increase the economic and ecological costs of farmers to a much higher extent than before.

Once again, this is being propagated in the name of increasing productivity but the reality is very different.

A group of eminent scientists from various countries who constitute the Independent Science Panel have said in their conclusion after examining all aspects of GM crops:

> GM crops have failed to deliver the promised benefits and are posing escalating problems on the farm. Transgenic contamination is now widely acknowledged to be unavoidable, and hence there can be no co-existence of GM and non-GM agriculture. Most important of all, GM crops have not been proven safe. On the contrary, sufficient evidence has emerged to raise serious safety concerns, that if ignored could result in irreversible damage to health and the environment. GM crops should be firmly rejected now.

Strong Economic Base Needed

Protecting Rural Decentralisation

ONE OF THE achievements about which India can rightly be proud is the vast network of three-tier local self-government institutions that has been created in all the over half a million villages of the country.

This system of 'panchayati raj' gets an important mention wherever rural decentralisation is discussed at the world level. About 2.8 million rural representatives are elected at regular intervals in this vast network of institutions of rural decentralization and over 1.2 million among them are women. Weaker sections including dalits and scheduled tribes are well represented.

The entire system, based on regular elections and continuity, is protected by constitutional sanctions (the 73rd amendment); if elections are not held in time then courts can be approached to ensure that elections for panchayats are held, as happened some time ago in Jharkhand.

Of course, in keeping with the increasing problems and corruption in the overall political system in the country, many vexing issues have also been increasing in the panchayati raj system and some aspects of the system have functioned at considerable variance from the intentions. At the same time, there are many examples of the welcome changes initiated by panchayati raj representatives, particularly women and weaker sections among them. If the overall performance of panchayats has been moderate, the future potential is much higher. If most problems can be sorted out, the creativity and cooperation of people for development tasks and environment protection can be harnessed much more effectively under the ageis of the panchayati raj.

Hence, the road ahead should be to strengthen panchayati raj while at the same time initiate some reforms on the basis of a careful appraisal of how the system of panchayati raj has worked so far. To give just one example of where reform is needed, it has been seen at many places that one person becomes all-important amongthe various team members, whether at the village, block or district levels.

So a process reforms to check such unwelcome hegemonic tendencies are certainly needed but the overall thrust should be to make an improved panchayati raj system stronger and increase its capacity to take on more responsibilities in keeping with its constitutional mandate as well as newly emerging factors that are further increasing the need for rural decentralisation. For example, climate change and increasingly erratic weather conditions have increased the need for decentralised decision-taking by rural communities to cope with fast changing situations.

Unfortunately, the changes that have been initiated in recent times have weakened panchayati raj instead of improving and strengthening it to meet the bigger challenges ahead. The gram sabhas or the wider assembly of all adults in the village communities have been provided some important rights concerning matters of wider importance for their village's welfare and these are particularly strong in scheduled areas, as per the provisions of the special panchayat law enacted for the scheduled areas (called PESA in short) and particularly keeping in view the protection of the rights of tribal communities.

In the Name of Development

However, with the recent emphasis on clearing more and more projects rapidly overruling environmental, social and other considerations there is an increasing drift towards ignoring the concerns voiced by gram sabhas and panchayats.The well justified democratic rights provided to gram sabhas and panchayats that are essential for the success of decentralisation are increasingly seen as needless, avoidable impediments in the rapid and arbitrary expansion of an over-centralised and inequality-based development model.

It is important to stand up for defending genuine decentralisation. There have been some important movements for strengthening gram sabhas, including the one led by BD Sharma which have been inspired by Gandhiji and these will be a source of strength for protecting panchayati raj and sincere rural decentralisation in difficult times.

The tensions arising from the clash of rural decentralisation with over-centralising tendencies in the economy were evident earlier also but these have certainly increased in recent times.

These trends should be resisted and the great potential

of improved and reformed Panchayati Raj in playing a very creative and significant role in improving rural life and helping communities to face new emerging challenges like climate change should be emphasised.

Village Community

At a time when the spectre of unemployment is haunting more and more people in the increasingly uncertain times of globalisation, Mahatma Gandhi's message of swadeshi offers a comforting alternative to community-level initiatives for protecting sustainable livelihoods.

The concept of 'Swadeshi' has been wrongly interpreted by some in a narrow chauvinist way. In reality, this concept provides a model for secure and sustainable livelihoods close to home not only in the Indian context but for communities anywhere in the world.

Briefly and simply – 'swadeshi' means that in any community, those needs which can be met by local production, should be met in this way. There is no insistence or obstinacy about unrealistic, non-viable efforts to produce those things locally which cannot be produced at the local level. At the same time, big corporations should be discouraged from producing those goods and services for which local production is viable and, in fact, more appropriate. If they insist on intruding on this market, then the community should mobilise local opinion to ensure that the interests of local producers are not harmed. As long as big corporations (or other producers from outside the community) do not intrude in those areas which are best served by local production, there is no conflict with them.

For example, let us consider a community consisting of a cluster of say, fifty villages, and two or three small towns.

Their community representatives, ensuring a good presence of women, can get together and decide which consumer goods, producer goods and services are most suitable for local production. While taking this decision, they'll keep in view the existing skills and capacities of farmers, artisans, workers etc. as well as other potential skills that can be developed keeping in view the needs of increasing local livelihood opportunities (including those skills which had flourished earlier but have been lost due to the invasion of local production by big corporations - example handloom weavers and oil millers).

A beginning can be made by playing a game which is in fact so simple that primary school children can happily join in this. In this game, the first step is that we make a list of all the goods and services we use from morning to night in a day. Then from this list, we make a smaller list of those goods and services which can at least theoretically be obtained at the local level. Then, finally, we prepare a third, yet smaller, list of those goods and services, which from the point of view of practical considerations also can be produced locally. The community then concentrate on ensuring that the goods and services listed in the third category are actually produced at the local level.

Apart from providing livelihood opportunities to a significant number of people, this will also ensure that the quality of goods and services will be better. For example, the quality of locally produced bakery products is likely to be better – these will be delivered fresh and the use of harmful preservatives can be avoided.

In the case of producer goods, or goods which are used to produce other goods of direct consumption, farmers may find it very useful to produce good quality local seeds instead of depending on the uncertain supply from outside sources.

Similarly, they may like to substitute chemical fertilisers and pesticides obtained from outside sources with locally available organic alternatives and producing and supplying compost can be an economically sustaining activity. In the case of services, a village community may decide that it will like to have its own youths as village teachers and health workers. The village community may even decide to jointly bear the medical education expenses of two or more youths in the village on the condition that after their education they'll come back to serve the village community as doctors.

Apart from ensuring quality, it'll also be useful to meet more and more basic needs (for example, upto 90% of food needs, even after allowing the option of obtaining a few fruits, vegetables and spices from outside), from within the community so as to increase the self-reliance of the community. The community will then be in a better position to face the sudden ups and downs of the global economy, compared to another community which depends on outside suppliers for meeting most basic needs.

Thus the Gandhian, swadeshi way of protecting secure and sustainable livelihood can also ensure the supply of quality goods and services, apart from making the community more self-reliant.

Protecting Livelihoods and Skills

In recent years, most reports about handloom weavers have spoken of decline and decay, debts and distress.

Yet it is possible to envisage steps and solutions which can not only protect lakhs of skilled livelihoods in handlooms and khadi, in addition it should be possible to create lakhs of new livelihoods – full-time and part-time – in this sector.

Handloom cloth is cloth woven by hand, or cloth woven on

manually operated looms (as distinct from electricity operated looms or powerlooms). Khadi is cloth which is on the one hand hand-woven (handloom) and in addition the yarn for weaving this cloth is also obtained by hand-spinning (for example on charkha or manually operated spinning wheel). Thus, khadi is hand-woven plus hand-spun cloth.

Khadi has a very special significance in India as during the freedom movement, Mahatma Gandhi gave great importance to khadi as a symbol of India's liberation (earlier India's weavers had been ruined by unjust promotion of British mill-made cloth by colonial rulers).

Present and Potential Employment

According to official data given by Development Commissioner (handlooms) of the Government of India, as per statistics available in the handloom sector in 2007, this sector is next to agriculture in providing employment to about 6.5 million persons.

The Tenth Five-Year Plan document 2002-2007 says:

> Handlooms are a part of India's rich heritage and exemplify the country's diversity and the artistry of the weavers. It plays a very important role in the economy. This sector is estimated to have provided direct and indirect employment to about 2.5 million households and about 12.4 million weavers and others engaged in weaving and allied activities in 2001-02. This sector contributes nearly 19 per cent of the total cloth produced in the country.

Further, this document talks about the problems of this sector, saying:

> The handloom sector is facing a number of problems like obsolete technology and traditional production techniques, high price of hank yarn, inadequate availability of inputs

> like standardised dyes and chemicals in small packs, lack of new designs, inadequate training for upgradation of skills etc. and inadequate marketing intelligence and feedback. Besides, it suffers from disadvantages like unorganised structure, weak financial base of the weavers and bureaucritisation/politicisation of cooperatives.

What this routine description of problems does not state is that in many belts of handloom weaving, the crisis of weavers has reached such a situation that very significant numbers of weavers have been forced to give up their traditional skills and take up employment as construction workers, rickshaw-pullers, vendors etc. (in Benaras, for example.)

The review of khadi cloth production in the Tenth Plan document also reaches a dismal conclusion, 'over the years, the production of khadi cloth has been on the decline. Hence employment in this area is also falling'. The reasons highlighted for the poor performance of khadi by this document include uncertainty over the continuation of the rebate policy for khadi, high stock of unsold khadi and mismatch in khadi production.

In fact, if we review the performance of khadi and handloom sectors, as well as related artisan skills, then there are several examples of how non-implementation of existing protective provisions led to large-scale loss of livelihoods.

As noted earlier, an expert on handlooms, LC Jain estimated (in 1983) that during the previous decade, that is in the '70s, 5.5 million handloom workers were rendered unemployed or their employment had been reduced due to the displacement of 1,386 thousand handlooms (each handloom providing part or full employment to 4 person) by 231,000 powerlooms (each powerloom displaced six handlooms).

Similar massive loss of employment was seen in related areas such as the hand-printing industry. According to Jain, because machines were employed for 942 million meters of clothes over and above the 500 million meters at which their output would have been frozen, as recommended by the Research Advisory Panel (textile printing industry) an estimated 2,50,000 jobs opportunities have been lost in the economy.

So, the situation today is what we see after several decadess of loss of livelihood opportunities. Hence the actual employment available in handloom and khadi today is much less than the potential. On the one hand, lakhs of existing livelihoods can still be saved. On the other hand, lakhs of livelihoods can also be added if the problems which caused distress to weavers can be removed, as they still have their skills and can return to livelihood based on their traditional skills if given the opportunities. If the outlook for handloom and khadi is good, many artisans will also feel encouraged to teach family skills to the next generation. Hence this livelihood will continue to be nurtured.

Inherent Strength

For many people who like to discourage such efforts, this is unrealistic as the inevitable march of mechanisation is bound to reduce the prospects for handloom and khadi. However, in an expanding market of cloth production, higher production by the mechanised sector can co-exist with higher production by khadi and handlooms if the potential of the later is allowed to be properly realised. What is important to emphasise is that handloom and khadi have some inherent strengths and some types of cloth is best woven only on handlooms.

In a review of the inherent strengths of handlooms, BK Sinha (Development Commissioner, Handlooms) pointed out that due to manual operations several combinations are possible in handlooms with intricate designs.

> 'The functional properties like drape, texture, strength, wrinkle resistance, dominant stability etc. can be ingeniously manipulated through appropriate designs, exclusive types of fabrics used, counts and twists of warps and yarns, thick density, type of weave, type of fashion and process employed in printing.'

This review goes on to detail many kinds of clothes which are best woven on handlooms:

> The clothes made from extremely fine material, i.e. yarn count with 100s and above which are delicate, can be woven more safely on the handloom owing to comparative lightness of jerks. The polish of the clothes interwoven with gold or silver thread, can be taken out by extremely frictional action of powerlooms. On the contrary, handlooms are ideally suited for such work. Clothes with multi- coloured designs in which the weft is to be changed very frequently are most suited to handlooms. Clothes with embellishment in the border and heading and entire body with delicate designs in various colours which calls for individual schemes can be ideally woven on handlooms. Short pieces of clothes having unique designs to meet individual tastes, which may not be economical for being produced on powerlooms/mills can be made by warp of short length with preferential colour on handlooms. Rough clothes of very low counts such as durries, floor coverings, rugs, etc. in which the tensile strength of the yarn is too low for the powerloom, can be done only through handlooms. Handloom has monopolistic position in the manufacture of checked and striped colour fabrics with numerous design and different weft colour, colour with fanciful yarns, durries, floor coverings, tapestries, etc. The necessity of using multiple box looms for giving more colour in the weft involves heavy cost in powerloom. Also insertion of multicoloured weft in different places, at comparatively

> short regular interval is difficult in powerlooms and mills, as it interferes in normal speed of loom. Handlooms are capable of producing most fanciful colours and designs in small quantities. Its flexibility of catering to very narrowly segmented market is advantageous to handlooms. (Quoted from *Yojna*).

Many people who have been using khadi cloth for years say emphatically it is very healthy for skin, providing comfort in summer as well as in winter. With growing health consciousness, this can help to increase the demand for khadi in India as well as abroad. Also, the undisputed fact that handloom and even more so khadi generates the most employment per metre of cloth can also be used to promote this cloth.

The consciousness for energy-conservation as well as reducing greenhouse gas emissions is increasing. From this point of view, hand-woven cloth has an obvious edge, particularly when it is also hand-spun. The eco-worth of these products can be increased further by the increasing use of vegetable colours.

Massive amounts of money are spent on imparting vocational education, including industrial skills in an institutional set-up. But in the informal set-up of handlooms and several related crafts and artisan skills, an informal structure exists for imparting invaluable and intricate skills to the next generation without the government spending any money. Surely such traditional sustainable livelihoods based on beautiful skills need to be protected and promoted in a big way.

What Needs to be Done

Probably due to the impact of the Gandhian legacy inherited from the time of the freedom movement after independence the government set in place a series of measures for the protection and promotion of handlooms and khadi. For example, spinning mills were obliged to provide a part of production of yarn as hank yarn to handlooms at a fair price. Certain types of cloth was reserved for handlooms. But gradually, these legislative measures were diluted and even what remained on paper was not implemented properly. It is important to strengthen these laws, widen their scope and above all to ensure their better implementation.

Mobilisation and organisation of weavers and related artisans can play a big role in creating conducive conditions for better implementation of protective laws and promotional schemes. In recent times as the crisis of weavers worsened in Varanasi and neighbouring districts, a well-organised effort by three organisations Banaras Bunkar Samiti, Human Welfare Association and Find Your Feet helped to draw widespread attention to their problems as well as take up the problems with government authorities.

While protecting handloom weavers is important, the cause of hand-spinning is no less important. Somehow the schemes of handloom promotion have got delinked from hand-spinning, although Gandhi had emphasised that without hand-spinning as a base, handloom weavers will never be able to ensure self reliant, sustainable development. They will remain dependent on mills for yarn, on suppliers who want the rapid development of mechanised weaving (apart from spinning).

Although an infrastructure has been created for the promotion of khadi and several schemes are in place, the condition of hand-spinners (mostly women) is precarious, they are paid poorly and there are reports of corruption spoiling the good name of khadi institutions too. So, clearly, much more remains to be done before the undoubted huge potential of protecting and promoting the livelihoods and skills of lakhs –

perhaps millions of artisans – can be achieved.

There are several struggles in India to give a fair deal to handloom weavers. Now, with the increasing emphasis on reducing GHG emissions, the case for protecting handlooms and hand-spinning becomes stronger as protectIon of precious human skills which do not involve the use of any commercial energy, fossil fuel use or the related GHG emissions has acquired a new strength, it can get more support than before. A fairtrade clothing company People Tree (Dhaka, Bangladesh) says in its promotional campaign that the handloom saves one tonne of CO2 from being emitted each year and creates nine times as many jobs as the the powerloom. (see *Resurgence* Jan-Feb 2010).

At a time when the need to reduce GHG emissions is so pronounced, technologies which do not involve any GHG emissions to create products of great value and utility should obviously be helped and protected, particularly in a country like India where despite mounting threats to handloom weavers and other artisans they have still survived in significant numbers and hence can provide the base for a wider revival of these skills and technologies.

Spiritualism and Spinning

To the criticism that khadi provided coarser and hence inferior cloth, Mahatma Gandhi replied that we must learn to appreciate the art associated with hand-made products.

He wrote:

> All the evidence hitherto collected goes to show that spinning is an elegant art and that the process itself is extremely pleasant. No mechanical pull is enough to draw the various counts. And those who do spinning as an art

> know the pleasure they derive when the fingers and the eyes infallibly guide the required count.
>
> True art can only be expressed not through inanimate power-driven machinery designed for mass production, but only through the delicate living touch of the hands of men and women (1929).
>
> The sacred touch of her hand gives life and history to khadi which the machine-made yarn can never give. The art that is in the machine-made article appeals only to the eye; the art in khadi appeals first to the heart and then to the eye (1934).

In fact, Gandhi went a step further and explained how spinning work could be raised to a spiritual experience. Describing an experience in 1925, when some musicians played for him while he was spinning, he wrote:

> I always use a noiseless charkha. It, therefore, did not interfere with my enjoyment of the music. On the contrary, it enhanced the pleasure of listening to the music, and the music enhanced the pleasure of spinning. And neither interrupted my communion with God. The hand, the ear and the heart acted in perfect harmony. Let the sceptics test the experience for themselves.

PEACE IS ITS OWN REWARD

I believe that Gandhi's views were the most enlightened of all the political men of our time. We should strive to do things in his spirit: not to use violence in fighting for our cause, but by non-participation in anything you believe is evil.

—Albert Einstein

The choice is no longer between violence and non-violence. It is now either non-violence or non-existence.

—Martin Luther King

I and others may be revolutionaries, but we are the disciples of Mahatma Gandhi, directly or indirectly, nothing more nothing less.

—Ho Chi Minh

Increasing Relevance

Peace Needed more than Ever Before

PROMOTING WORLD PEACE and avoiding all wars have always been very important (although seldom realised) objectives for humanity, but in recent decades their importance has increased due to three reasons. Firstly, the availability of weapons of mass destruction (WMDs) has meant that the destructive potential of war is many times more than ever before. In an all-out war using WMDs, most life-forms on earth will be destroyed in a short time, even though all other life forms and over 95 per cent of human beings are not at all responsible for such a war and its weapons. This situation is unique to recent decades, even though life on earth has existed for billions of years and human life in its present form has existed for over 40,000 years at least.

Secondly, a number of very serious inter-related environmental problems, perhaps led by but certainly not confined to climate change have emerged in recent decades, quite capable of destroying or seriously disrupting the basic life-nurturing conditions of our planet. In order to resolve these problems before it is too late, we need continuous peace in the next few decades, so that the much-needed effort and resources can be devoted for resolving the serious environmental problems without being encumbered or disrupted by wars and related tensions.

Thirdly, wars, weapons and war-preparations, creating and maintaining the related infrastructure, are themselves a huge source of environmental ruin including greenhouse gas (GHG) emissions. It is within human capacity to reduce this source of environmental ruin by about 95% or so in a short time, also releasing human and material and financial resources for much-needed protective tasks.

Thus, while world peace has always been a very desirable objective, this has been particularly so in recent times. Yet, the present-day world leadership has not been able to make any significant progress towards ensuring peace in recent times. On the contrary, there are all the signs of moving in the reverse direction, including increasing hostility among some of the most powerful countries and the non-renewal of whatever limited agreements that had been reached earlier to reduce the high risks of WMDs.

This failure of world leadership (persons and institutions) at a very critical stage of human and Earth's history can have catastrophic consequences. On the one hand, there is high risk of very destructive wars and on the other hand, even if there is no actual war, war-tensions and war preparations can create a situation in which it is not possible to make continuing efforts with cooperation of all nations to resolve the major environmental issues including climate change before it is too late for humanity.

In the middle of this darkness, where is light? Where is hope?

One possible source of hope is a great resurgence among ordinary folk, among common people, among most of us for peace. But how is this possible? One answer is to link world peace to peace in daily life of people. Another and related answer is to link the wider peace objective closely to justice for people at all levels – economic and social justice including gender justice.

In this context the work and ideas of Mahatma Gandhi in particular are very relevant as for nearly five decades he worked for resisting injustice at various levels in peaceful ways, experimenting with various forms of peaceful resistance and evolving them all the time and also taking this message of hope to millions of people.

He succeeded in convincing millions of people to embrace peaceful methods for obtaining justice for several decades with continuity, which in itself is a very important achievement.

This has left a lasting legacy as many struggles in the world including those led by Martin Luther King and Nelson Mandela were inspired by these struggles and experiments of Gandhiji. These struggles in turn inspired others.

Even great revolutionaries who in their own circumstances chose different paths paid a tribute to this messenger of peace. Ho Chi Minh, the great revolutionary leader of Viet Nam, declared himself to be a disciple of Mahatma Gandhi. Subhash Chandra Bose, the valiant and great freedom fighter who in certain circumstance had left the path of Mahatma Gandhi, later addressed him as the 'Father of the Nation' and gave him a new title that has persisted for so long amidst many odds.

Despite all this, some limitations of what was achieved also became apparent with the passage of time.

India's freedom was accompanied by the most horrific violence related to partition, despite the untiring efforts of Gandhiji and his colleagues to prevent this. This is stated not as a criticism, but in all sincerity to learn from history so that even a more broad-based approach to peace can be explored in future as the challenges ahead of us are truly huge ones.

The world today needs a peace movement with continuity which links up peace in everyday life with the big issues of world peace. When the peace movement helps people in many ways in their daily life, they join the peace movement in larger numbers. All the time the peace movement makes efforts to make people aware of the big issues of world peace, including eliminating WMDs and wars. Hence, the support for the big issues of world peace builds up at the grassroots. Let us consider two theatres of long-drawn wars, Afghanistan closer at home and Iraq, the cradle of middle-eastern civilisation in Gandhian terms. From this flows the thirst for world peace.

Governments and international institutions have to respond to this great upsurge in people's yearning for peace. At the same time, the peace movement needs to work with selected and more responsive persons in governments and international institutions to elicit a better response from them.

In all this we have to learn much from Mahatma Gandhi, just as we have to learn much from his several colleagues as well. This includes well-known colleagues such as Badshah Khan (also known simply as Frontier Gandhi) as well as lesser known followers (like Sridev Suman of Uttarakhand) who made great sacrifices on the path of peace. Similarly, we need to learn from great teachers like Tolstoy, who inspired Gandhiji, and from Nelson Mandela and Martin Luther King, who were inspired by Gandhi.

We should also learn from the experiences of the various social and environmental movements (the Chipko and Appiko movements to save forests in India, the Narmada Bachao Andolan etc.), various peaceful land justice struggles as well as the even more numerous peaceful anti-liquor struggles with a strong support-base of women, which were inspired at least to some extent from Gandhiji's ideas and efforts. There is an increasing relevance of various efforts to check inter-faith violence and to strive for inter-faith harmony, a cause which draws often from the views and commitments of Gandhiji.

Lessons from Taliban Land
Forgotten History of Non-violent Struggles

IN RECENT YEARS the spread of violence in the Pakhtun-dominated area of Pakistan with the growing strength of the Taliban in these areas has been a matter of great concern.

At the same time, it is important to recall that under the inspiring leadership of Khan Abdul Ghaffar Khan (also called Badshah Khan - the King of Khans - and Frontier Gandhi) a highly inspiring and completely non-violent freedom movement had prospered in exactly the same geographical region during 1920-1947, peaking around 1930-1935 with the involvement of nearly a hundred thousand soldiers of peace. Badshah Khan was very close to Mahatma Gandhi and both of them felt deeply inspired by the work of each other.

Badshah Khan had successfully raised an 'army' of peaceful 'soldiers' called *Khudai Khidmatgar*s (or servants of God).

Every Khudai Khidmatgar had to take this oath:

> I am a Khudai Khidmatgar; and as God needs no service, but serving his creation is serving him, I promise to serve humanity in the name of God.
> I promise to refrain from violence and from taking revenge. I promise to forgive those who oppress me or treat me with cruelty. I promise to refrain from taking part in feuds and quarrels and from creating enmity. I promise to treat every *Pathan* as my brother and friend. I promise to refrain from anti social customs and practices. I promise to live a simple life, to practice virtue and to refrain from evil.
> I promise to practice good manners and good behavior and not to lead a life of idleness. I promise to devote at least two hours a day to social work.

Describing the importance of this oath for the people of this region, Badshah Khan's biographer Eknath Easwaran writes:

> For a Pathan, an oath is not a small matter. He does not enter into a vow easily because once given, a Pathan's word cannot be broken. Even his enemy can count on him to keep his word at the risk of his own life. Nonviolence was the heart of the oath and of the organisation. It was directed not only against the violence of British rule but against the pervasive violence of Pathan life.

During 1930-31 came the real test of edurance of their commitment to non-violence as the colonial police and army rapidly escalated their brutal repression to check the spread of the freedom movement.

According to a report prepared by a Congress Inquiry Committee, a protest demonstration in Qissa Khawani Bazaar was dispersing peacefully when:

...all of a sudden, two or three armored cars came at great speed from behind without giving warning of their approach and drove into the crowd. Several people were run over, of whom some were injured, and a few killed on the spot. The people were not armed - [not even with] stones or bricks. The crowd behaved with great restarint, collecting the wounded and dead.

More people collected. The troops were ordered to fire. 'Several people were killed and wounded', the report continues:

> and the crowd was pushed back some distance. At about half past eleven, endeavours were made by one or two outsiders to persuade the crowd to disperse and the authorities to remove the troops and the armored cars. The crowds were willing to disperse if they were allowed to remove the dead and the injured and if the armored cars and the troops were removed. The authorities, on the other hand, expressed their determination not to remove the armored cars and troops. The result was that the people did not disperse and were prepared to receive the bullets and lay down their lives. The second firing then began and, off and on, lasted for more than three hours.

In his study of nonviolent movements, Gene Sharp of Harvard includes a description of the firing in Qissa Khawani Bazaar:

> When those in front fell down wounded by the shots, those behind came forward with their breasts bared and exposed themselves to the fire, so much so that some people got as many as 21 bullet wounds in their bodies, and all the people stood their ground without getting into a panic. A young Sikh boy came and stood in front of a soldier and asked him to fire at him which the soldier unhesitatingly did, killing him. The crowd kept standing at the spot facing the soldiers and were fired at from time to time, untill there were heaps of wounded and dying lying about.

A newspaper of Lahore, which represented the official view, itself wrote to the effect that the people came forward one after another to face the firing and when they fell wounded they were dragged back and others came forward to be shot at. This state of things continued from 11 till 5 in the evening.

Such incidents were repeated. The brave Pathans continued to pass the test of endurance and their commitment to peace and non-violence. The result was that within a few months their support increased dramatically. The Khudai Khidmatgars had started with a strength of about one thousand only, but by the end of the September that year, there were nearly eighty thousand volunteers.

Another equally inspiring and dramatic development was that the Hindu soldiers of Garhwal Rifles refused to fire on peaceful Pashtun freedom-fighters, even when they knew that they'll get the most severe punishment. In fact, these brave soldiers led by Chandra Singh Garhwali told their officers that they could blow them from their guns if they wanted to but the soldiers would not fire on freedom fighters.

The peaceful resistance spread widely despite the fact that Badshah Khan and other leaders were arrested. His two sons and elder brother were also arrested.

The Pathan Women

Inspiring aspect of the movement very relevant to the present days of the Taliban is that Pathan women were encouraged to participate actively in the freedom movement and many of them came forward to play an important role.

Badshah Khan opposed the purdah system (veil) and emphasised the equality of women.

Consider the example of a man called Murtaza, who had earlier been jailed for murder, who came under the influence of this peaceful freedom movement.

Although he could not sustain this non-violence throughout his life, at least for some years he observed it honestly and sincerely. When Badshah Khan's son Ghani interviewed him about this period of his life, Murtaza replied, 'I was a little saint for those four years. I tried to live up to my dreams instead of my desires. It was great, it was a miracle. I refused fortunes for a hope and spared lovely girls because they trusted me and looked up to me. You cannot help loving those that love you, and you cannot hurt those that trust you. I tried to live up to what the people thought I was'.

Lessons from the Bygone Era

The lessons that one can learn from recalling this glorious phase of non-violent struggles are three.

Firstly, the entire Pashtun community should not be given a bad name because of the spread of Taliban in the area. Till as late as 2007-08, they voted for and elected secular political parties. Many of them resisted the spread of Taliban even at the cost of their life.

Secondly, this phase of history tells us how the undoubted courage of people who live violent lives full of 'honour' and 'revenge' killings can be harnessed for constructive tasks and struggles for justice.

It is when such efforts are constantly suppressed (Badshah Khan and his followers faced jails and repression even at the hands of Pakistan's government after the end of the British rule) that the chances for feeling of oppression and injustice to be channelised into the path of fundamentalism and religious extremism increase.

This is particularly true if the government itself encourages this as has certainly happened in Pakistan for a long time.

Revival of Khudai Khidmatgars in India

Keeping in view the great importance of the revival of Khudai Khidmatgars in the entire history of non-violent struggles against injustice as well as service of humanity, it is a highly welcome news that this organisation has been revived in India sometime ago.

Faisal Khan, National Convenor of Khudai Khidmatgars in India, says that efforts were being made for quite sometime, but the formal inaugural meeting took place near the martyrdom site of Mahatma Gandhi in Delhi on 20 January 2011. Following this, the immediate membership drive proved to be successful even in more remote States like Assam and Manipur.

At the follow-up meeting held at the same place on 20 January 2012, several uniformed Khudai Khidmatgars were also present who have special responsibilities for serving in areas which suffer from communal strife and natural disasters. Leading Gandhians expressed their hopes and good wishes for the steady progress of Khudai Khidmatgars in India. Today there are more than ten thousand members of the organisation in India. The movement can be revived in its birthplace too, if there is a political will to support this.

Gandhi in Baghdad
Imagining a Resistance

MAHATMA GANDHI'S VIEW of the valiant raises the question – can't we have a different vision of resisting military might? Can't we think of resisting or even stopping war by peaceful means?

Let's try to imagine how peaceful resistance would've looked like in Baghdad (from 2003 to 2011) under the leadership of Mahatama Gandhi.

As the invading army entered, it would've been greeted by peaceful, well-organised Iraqi men, women and children – particularly women – gathered in large numbers and carrying banners which proclaimed: 'Please go back', 'We will decide our own future', 'No return to colonial ways', 'Foreign army can't bring democracy'.

This Iraqi peaceful resistance would've been accompanied by hundreds, perhaps thousands, of volunteers from other countries, with special emphasis on invader countries.

Wherever invading soldiers went, they would be 'greeted' by these peaceful protesters. Local collaborators of invader army would also face similar resistance.

If the invaders still did not go back, the people would've declared a movement of complete non-cooperation with the invader-led administration, exempting only relief and medical work. Peaceful resistance would be offered at various levels to invader-led administration.

This peaceful resistance would be supported in a big way by peace-loving people all over the world, particularly in invader countries, who would organise several supportive actions such as relief and publicity. This resistance would have continued till the invaders agreed to go back.

In this context, one must recall that the invader has had previous experience of non-violent movement; the American Civil Rights Movement led by Martin Luther King was one of the biggest anti-aparthied movements in the world not so long ago (the 1960s).

However, to be possible and successful, such peaceful resistance requires broad-based unity of people, a lot of preparation and training, a firm faith in non-violence and above all, tremendous courage. As Mahatma Gandhi said, 'Complete non-violence requires the highest form of courage'.

Cry also for the conquerer

Invading soldiers also suffer. The first priority of the peace-loving people should of course be to condemn any invasion like the American/British invasion of Iraq in the strongest possible terms. But, at the same time, care should be taken to ensure that this protest should not take the form of hatred. Even when we speak against policies of invading countries based on greed and dominance, care should be taken not to condemn the ordinary people of the conquerer socities (who actually may themselves be victims of such policies).

The soldiers of the invading armies are given very cruel tasks by their governments. Beneath their helmets and uniforms, they too are human beings with fears and feelings. These long years of war has affected the victors in two different ways. Firstly, there is the risk of personal injury.

Secondly, there is the less visible but nevertheless perhaps even more significant risk of how their minds and hearts have been affected on a permanent basis when they have performed the very cruel and unjust responsibilities that have been assigned by their governments.

After all, these young men and women have absorbed at least some ideals in their education. Before being sent to Iraq and Afghanistan, they were encouraged to think of themselves as *liberators.* When they see that their actual responsibility is highly unethical and cruel, this is bound to have a disturbing impact on the minds of these young soldiers of the invading army.

Speaking of their experience of two earlier wars, the veterans of the US armed forces said in a statement at that time:

> In the last Gulf War, as troops, we were ordered to murder from a safe distance. We destroyed much of Iraq from the air, killing hundreds of thousands, including civilians. We remember the road to Basra – The Highway to Death – where we were ordered to kill fleeing Iraqis. We bulldozed trenches, burying people alive. The use of depleted uranium weapons left the battlefields radioactive. Massive use of pesticides, experimental drugs, burning chemical

> weapons depots and oil fires combined to create a toxic cocktail affecting both the Iraqi people and Gulf War veterans today.

Thus, one in four Gulf War veterans is disabled.

> During the Vietnam War we were ordered to destroy Vietnam from the air and on the ground. At My Lai, we massacred over 500 women, children and old men.
> This was not an aberration; it's how we fought the war. We used Agent Orange on the enemy and then experienced first hand, its effects.
> We know what Post Traumatic Stress Disorder (PTS) looks, feels and tastes like because the ghosts of over two million men, women and children still haunt our dreams. More of us took our own lives after returning home than died in battle.

It is evident from this statement of war veterans, who have seen and experienced present-day military invasions more than anyone else, that:

- So dangerous are present-day weapons that even the winning side of a one-sided war is likely to suffer heavy physical damage.
- Even in one-sided wars, soldiers of the stronger side are habitually asked by superiors to inflict massive unbearable cruelty and kill without any need for killing.
- All this cruelty is of course terrible for the victims, but it also leaves permanent scars on the victors, on soldiers who are forced to inflict these cruelties.

This last aspect, frequently ignored, is very important *for understanding the enormous costs of war for the victors.*

As the veterans of US armed forces have clearly said of the Vietnam experience, 'More of us took our own lives after returning home than died in battle'.

The parents of many American soldiers going to war are aware of these risks, as is evident in several statements released by the organisation, Military Families Speak Out. One parent, Stephen Cleghorn said at the time of the Iraq war, 'If we go to war in Iraq, the loss of innocent civilian lives will be high and horrific.... The soldiers in the area will know what they have done. They will see it with their own eyes, or they will see it in the eyes of their fellow soldiers'.

Nancy Lessin and Charley Richardson said about their son, 'We don't want him to be wounded or die. We don't want him to be forced to wound or kill innocent Iraqi civilians. That would kill a part of him – and a part of us'.

This then, is the invisible cost of war.

Reader's Digest has described the suffering of a child, Kim Phuc, at the time of bombing of Trang Bang (Vietnam) by US planes (RD November 1997):

> The bombs, canisters filled with napalm, had smashed into ground behind Kim and instantly ignited. The jellified gasoline, designed to stick to and incinerate anything it touches, splashed on to Kim's back. Her flowered cotton shirt and pants – even her sandal – combusted. She was engulfed in a cloud of smoke and fire as napalm peeled away the skin from her back and left arm. Terrified, Kim kept running. At first she could feel nothing. Then she felt as if she had been thrown on to an open fire. In horror she saw the skin drop off her arm like clothes off a doll. As she ran naked down the road that led out of the village, she began screaming, 'Too hot! Too hot! Please help!'

This magazine also noted the impact of this suffering on the sensitive mind of the pilot who caused this suffering:

> Now he stared at the picture of Kim Phuc, her agony caught for eternity. His own son Louis was about the same age. He could almost smell the child's burning flesh.

> Later he kept his role in the bombing of Trang Bang secret, locked deep within his soul.
> It surfaced in the form of a nightmare. First Plummer (the bomber pilot) would see a picture of Kim, with arms outstretched and mouth frozen in a silent scream. Then the image would widen to include Kim's brother and cousins running alongside her. Finally, he would hear their screams, louder and louder until he felt surrounded by the accusing children.
> To drown his guilt, Plummer began drinking heavily. In July 1973 he married for the second time, but he still kept his secret. No one can understand, he thought. John Plummer's drinking cost him his marriage in 1979. It was a vicious circle; he drank to put the bombing out of his mind, but the drinking made him more obsessed.

Apart from tormenting the sensitive mind forever, wartime cruelties can also have a somewhat different impact. In order to come to terms with the cruelties inflicted by them, some soldiers deliberately train their minds to become very insensitive to human suffering. This insensitivity later enters into their close personal relationships and can destroy them.

Time magazine has reported that spousal abuse is occurring in 1 out of every 3 US Army families each year – double the civilian rate. Some cases of violence at the hands of husbands (or other men in family) are so gory that not many people will be able to stand any detailed descriptions.

As reported in *Time*, a marine in California, who was a 'hero' of the Persian Gulf war, not only killed his newly-divorced wife but in addition also killed their five-year-old daughter. In North Carolina, an airman hacked his wife to pieces, wrapped her remains in plastic garbage bags and stored them in the refrigerator.

As I've concluded in my book, *Another Path Exists:*

> A person (or group or nation) who inflicts injustice and injury on others, will either live with a guilt complex (if he wants to retain some sensitivity) or else he'll have to reduce himself to a level of insensitivity that will prevent him from feeling small but precious joys of everyday life and this, in turn, is bound to adversely affect his closest relationships including those with his family members. Thus, relationships of dominance are not only destructive, these are also self-destructive.

Burning at Both Ends
Understanding Critical Links

WARS DESTROY NOT only the vanquished but also the victors, exploitation shatters not only the exploited but also the exploiters.

The wealth obtained by a person or society cannot be separated from the methods used for obtaining this wealth. If the methods are unfair, if these involve the unfair deprivation of others (including future generations), then these methods also have in them the seeds of guilt that will prevent the person or the society from deriving real happiness on the basis of this wealth.

The body and soul comprise one whole. One cannot be separated from the other. The material and spiritual aspects are closely related. The two cannot be separated. The tendency to curb one's spiritual life and indulge in unjust acts to acquire wealth does not bring the expected happiness. The suppression of spiritual values (that is required to be able to commit injustice) creates distortions in those persons due to which they are denied even that happiness which many poor people can experience freely and in plenty.

The persons who fill their coffers by inflicting injustice and cruelty on others have to live with a guilt complex that can destroy their peace. To get rid of this guilt, they have to lower themselves to such a level of insensitivity that deprives them of simple yet precious joys.

In ancient history, the later stages of Roman Empire perhaps represent the peak of the tendency to deprive others for the pursuit of one's pleasure. While a large number of slaves toiled in intolerably exploitative conditions, the Roman elites went to incredible extent in the pursuit of sensual pleasures. The pleasures they sought grew increasingly distorted, culminating in the monstrosity of gladiator fights. This was the time when the Roman empire was at the peak of its power. Yet, the life of the elite was completely hollow. While a part of their effort was concentrated on keeping a tight grip on their slaves (including brutally suppressing the revolts of the slaves such as the one led by Spartacus), so that they could have access to all the sensual pleasures without doing much work, this entire effort had created so many distortions in their life that they gradually failed to experience such happy experiences as come from genuinely warm human relationships.

What is more, the seeds of destruction of their apparently invincible power were also sown in this trend of having an empire based on the exploitation of others.

Not Much Different

Present times also see the culmination of a much bigger, almost worldwide trend of imperialism that started about five hundred years ago. The efforts of a relatively small number of people (the imperialists) to loot a much larger number of people (the colonised), to enslave them, caused famine and disease on a scale never seen before. This phase is still continuing.

But the ruthless plundering of other people in the years of colonialism and slave-trade, and the persistence of this in our own times of neo-colonialism and 'structural adjustments' has not been without its baneful impact on the exploiters themselves. In fact, in more ways than one, it can be seen that the economic deprivation seen in poor societies and the moral decay and social disintegration seen in the exploiter societies are the two sides of the same human tragedy.

The basically exploitative attitude towards the colonised and all the cruelties that emanated from this attitude, was responsible to a large extent for creating a spiritual void among the people and leaders of the coloniser societies which was responsible for much of their distress.

From the days of the slave-trades and the massacres of native Americans to the present day, even more destructive bombing of Vietnam and Iraq, what the history of the conquerer countries shows over the last almost 500 years is a constant willingness to use extremely cruel and unethical means to advance their economic and political interests.

This willingness cannot be reconciled with what has to be spoken at the political platform, what has to be said in the church, what has to be taught in the schools, what has to be written in popular press. *So, there is an unavoidable tendency to speak lies whether at the level of politics, religion, education or mass media. The people of nations which have to constantly lie to themselves about their role in the world cannot avoid becoming the victims of a serious spiritual crisis.*

The group of people who exploit and plunder others for self enrichment not only deprive others; in the process of plunder they also deprive themselves. They create within them a spiritual void that ruins their lives. This is true not only of the conquerer countries but also of the developing countries. The elites of the other countries who plunder and exploit their own people also suffer a similar deprivation of their own spiritual life, manifesting in similar types of problems.

The act of plunder and exploitation changes a person's soul in such a way that he becomes incapable of feeling and experiencing many kinds of happiness.

Such a person tries to isolate himself as much as possible from the suffering caused by his actions, seeking also the help of high technology for this purpose, but such efforts are self-defeating. The biggest tragedy of this deprivation is that the loss is generally not realised by those who experience it, hence proper corrective action is seldom taken (unless strong efforts are made to increase their awareness). In fact, the persons (and society made up of such persons) passing through this phase of deprivation are generally quite heady and arrogant due to a wrong sense of their power and dominance. Relationships of exploitation, therefore, destroy not only the exploited but also the exploiters.

However, there is also hope as efforts to end the exploitative relationship will liberate not just the exploited but also the (former) exploiters.

The self-defeating nature of the exploitative relationship (and the related liberating nature of the efforts to end exploitation) can be seen not only at the level of bigger communities but also at the level of the smallest social unit – right within the family. Insincerity and sefishness can be covered up by gifts and sweet talk only for some time. Those who seek to derive comforts from relationships of dominance and sefishness do not realise that they are making themselves incapable of deriving some of the most invaluable moments of happiness that come from a deeper relationship with our loved ones.

A dominant characteristic of the exploiter is that he (it is generally *he* rather than she) doesn't want to be told about the self-defeating nature of such attitudes.

An aggressive husband is just likely to dismiss the arguments given above. At the wider level of nations, the USA government goes to the extent of invading other countries to counter the drugs problem but is unable to realise how the drug addiction problem in the US is rooted in its own mistakes. However, the efforts to bring such a realisation should continue as such a realisation on the part of members of exploiter societies is crucial to the creation of a better world.

The process of inflicting cruelties on other people for the sake of securing undue benefits for oneself causes grave harm also to those who inflict these cruelties. Such acts create within them a certain guilt, a certain disquiet, a certain escapist trait to avoid facing the reality – that adversely affects their life in several ways, including their relationships with their near and dear ones. The more they try to curb their sensitivity to the suffering of others, the more incapable they become of experiencing the finer aspects and sentiments of human life. Without realising this, they shut themselves off from some of the best experiences of human life and interactions. On televison they like to see mostly programmes based on sex, violence, greed and pursuit of sensual pleasures. Their vision becomes narrow with the result that they cannot come to terms with the many dimensions of life. Their life becomes spiritually barren. They suffer but do not know why they suffer. The sense of arrogance and power on the one hand and the inability to make their life meaningful on the other reduces them to leading a life without purpose and meaning. The lack of spiritual strength pushes them into seeing life merely in terms of material achievements, hence they move from arrogance to a feeling of total loss of confidence, depending on ups and downs of material achievements.

The World of Wars

Persons and societies who plunder others or try to inflict cruelties on others generally try to distance themselves from the suffering they cause. Historian Raul Hillberg has written in the context of the 'holocaust' which killed hundreds of thousands of Jews in Hitler's Germany:

> It must be kept in mind that most of the participants (of the holocaust) did not fire rifles at Jewish children or pour gas into gas chambers... Most bureaucrats composed memoranda, drew up blueprints, talked on telephone and participated in conferences. They could destroy a whole people by sitting at their desk.

Elaborating this theme further, Ravi Sundaram writes in his review article on Zygmunt Bauman's book, *Modernity and the Holocaust* (*Economic and Political Weekly*, 29 February 1992):

> The emergence of a complex division of labour under modern capitalism has meant that functional specialisation generates necessary remoteness of human agents from the end-product of their social action. In this context, the bureaucrat's own action becomes an end in itself. Once so isolated from the

> consequences of action, the bureaucrat, untroubled by moral dilemmas, can pursue his allocated tasks.
>
> The architects of the holocaust, the bombers of Hiroshima and Nagasaki, and the technocrats who designed the Vietnam war, could continue doing so without moral qualms precisely because of the social production of distance in modernity. This aspect is crucial in understanding technological evolution of the holocaust. In the early history of the holocaust, the victims were rounded up and machine-gunned at point-blank range. The administrator soon found this both primitive and inefficient and damaging to the soldier's morale. *Other techniques were sought which would preserve the optical distance between murderer and murdered.* The result – the gas chamber, the perfect murdering machine. This reduced the role of the killer to that of the 'sanitation officer' who simply pressed the button which released the gas into the chamber filled with the hapless victims.

Apart from employing high-tech methods to widen the distance between the cruel action and the sufferings caused by these, the perpetrators of cruelties also try their best to manipulate the media. As *The Pioneer* newspaper reported (17 January 1992):

> The Reagan and Bush administrations have manipulated war coverage so drastically that the American public has a distorted picture of conflicts from Grenada to the Gulf, according to a study released by the Centre for Public Integrity. Curbs on the press are so extensive that White House and Defence Department leaders are violating the Pentagon's own information guidelines, says the Centre, a non-profit organisation founded by former ABC and CBS television News Producer Charles Lewis to examine ethics-related issues in Washington.

But, of course, all these efforts do not really succeed in removing the impact of the actions of cruelty. One can turn one's eyes from the suffering caused by cruel acts. One can convince, or force, the people around not to speak of the suffering and the cruel acts which caused it. But one cannot prevent the invisible yet inevitable shadows of guilt to become so imposing that they overwhelm the carefully cultivated outward appearance. These enter into the closest relationships, and even when one shuts the door to all relationships, the guilty person has to reckon with his own soul.

By curbing dissent and controlling media, a guilty society can at best succeed in making the guilt more silent but thereby, even more difficult to bear. The hypocrisy and double standards of such a society can badly shatter its own peace.

Since the truth cannot be hidden fully, there is an uneasy guilt and the accompanying sullenness eating the innerds of the people who know but will not admit. People cannot face the facts; they cannot face the reality of the real world. Since human beings cannot be made to live in two worlds constantly, gradually the same attitude – to discard ethics and moral courage – spreads also to domestic and personal issues.

One attitude that prevailed in an earlier era of colonialism was for the imperialist powers to reach a certain level of understanding among themselves so as to be able to divide the colonies among themselves. However, since the seeds of greed and cruelty remained, it did not take long for the understanding to be replaced by mutually destructive and fierce competition which played an important role in starting the First World War.

The First World War, which claimed about nine million lives, gave an idea of how destructive an all-out war in the twentieth century can prove. Meanwhile, the colonies of the West were increasingly becoming restive and demanding freedom. Surely, the only rational course for the European nations from a purely selfish point of view would have been to mend the mutual differences and concentrate their energies on controlling the colonies. Much as they would have liked to do so, they could not do this. Their mutual hatred continued to grow and ultimately plunged the planet into II World War.

Why did this happen, despite the fact that the European nations prided themselves in being the most civilised and educated? An important part of the answer – this part has also been the most neglected – is likely to be that the entirely unprincipled and unethical greed and cruelty which had gone into the plunder of colonies (and the natives of the new world) had created such a crises of spiritual values that attitudes like greed, hatred and cruelty could not be controlled within the most powerful nations even when it became most important to curb them for the protection of their own people. The uncontrolled fires of greed and cruelty which first consumed the people of the colonies and natives, later consumed the exploiters themselves in two World Wars.

The world has been changing so rapidly in recent years that no one can predict with certainty the fast – changing relationships, alliances and hostilities. The friends of today can be the foes of tomorrow and vice-versa.

What is much more important – and this will benefit all people of all nations, including the USA – is to look at the structural causes of war and tension and try to remove these structural causes. If it is the industrial – military establishment of a few countries – then let's oppose it. If it is the basic attitude of greed, then let us try to fight this. But let us not run away from the basic issues. Remember, as long as the structural causes are not tackled, no one can be safe.

To summarise then, to make this world safe and peaceful in the true sense of the word, the basic attitudes of greed, cruelty and exploitation have to be confronted and changed, instead of being hidden under hypocrisy, as these have the power to ultimately destroy not just the exploited, but also the exploiters.

Even though the economic, the political, the social, the ecological crises in the world is accentuating at a very rapid pace, the world's leaders by and large refuse to acknowledge that there is something fundamentally wrong with the world.

If they talk of social justice at all, the suggestions they give are an insult to the basic changes that are needed. It is in this dismal situation that the above viewpoint – namely that social justice is essential to liberate not only the exploited but also the exploiters – should be seen. The increasing acceptance of this viewpoint, we believe, will help to create a public opinion that will not tolerate hypocrisy in the name of social justice, and thereby pave the way for real changes.

Opposition to Invasion
Byrd's Peace Bid

IN HIS LAST years, Tolstoy pleaded very strongly with soldiers and would-be-soldiers of invading countries not to remain part of (or join) any invading army and its destructive actions.

The larger context of this in present day world is that all forces of peace should form broad-based unity to resist invasion and war. If a sustained and broad-based effort is made and a potentially very destructive invasion can be stopped, then this will be a big achievement which can inspire other such efforts.

In this context attention may be drawn here to a passionate speech made by Senator Robert Byrd on the floor of the US senate just before the invasion of Iraq.This speech one must take note of:

> I believe in this beautiful country. I have studied its roots and gloried in the wisdom of its magnificent Constitution.

I have marvelled at the wisdom of its founders and framers. Generation after generation of Americans has understood the lofty ideals that underlie our great Republic.

I have been inspired by the story of their sacrifice and their strength.

But, today I weep for my country. I have watched the events of recent months with a heavy, heavy heart. No more is the image of America one of strong, yet benevolent peacekeeper. The image of America has changed.

Around the globe, our friends mistrust us, our word is disputed, our intentions are questioned.

Instead of reasoning with those with whom we disagree, we demand obedience or threaten recrimination.

Instead of isolating Saddam Hussein, we seem to have isolated ourselves. We proclaim a new doctrine of pre-emption which is understood by few and feared by many. We say that the United States has the right to turn its fire-power on any corner of the globe which might be suspect in the war on terrorism. We assert that right without the sanction of any international body.

As a result, the world has become a much more dangerous place.

We flaunt our superpower status with arrogance. We treat UN Security Council members like ingrates who offend our princely dignity by lifting their heads from the carpet. Valuable alliances are split.

After war has ended, the United States will have to rebuild much more than the country of Iraq. We will have to rebuild America's image around the globe.

The case this Administration tries to make to justify its fixation with war is tainted by charges of falsified documents and circumstantial evidence. We cannot convince the world of the necessity of this war for one simple reason. This is a war of choice.

There is no credible information to connect Saddam Hussein to 9/11. The twin towers fell because a worldwide terrorist group, Al-Qaida, with cells in over 60 nations, struck at our wealth and our influence by turning our own planes into missiles, one of which would likely have slammed into the dome of this beautiful Capitol except for the brave sacrifice of the passengers on board.

The brutality seen on September 11th and in other terrorist attacks we have witnessed around the globe are the violent and desperate efforts by extremists to stop the daily encroachment of Western values upon their cultures. That is what we fight. It is a force not confined to borders. It is a shadowy entity with many faces, many names, and many addresses.

But, this administration has directed all of the anger, fear, and grief which emerged from the ashes of the twin towers and the twisted metal of Pentagon towards a tangible villain, one we can see and hate and attack. And villain he is. But, he is the wrong villain. And this is the wrong war. If we attack Saddam Hussein, we will probably drive him from power. But, the zeal of our friends to assist our global war on terrorism may have already taken flight.

The general unease surrounding this war is not just due to 'orange alert'. There is a pervasive sense of rush and risk and too many questions unanswered. How long will we be in Iraq? What will be the cost? What is the ultimate mission? How great is the danger at home?

A pall has fallen over the Senate Chamber. We avoid our solemn duty to debate the one topic on the minds of all Americans, even while scores of thousands of our sons and daughters faithfully do their duty in Iraq.

What is happening to this country? When did we become a nation which ignores and berates our friends? When did we decide to risk undermining international order by adopting a radical and doctrinaire approach to using our awesome military might? How can we abandon diplomatic efforts when the turmoil in the world cries out for diplomacy?

Why can this President not seem to see that America's true power lies not in its will to intimidate; but in its ability to inspire?

> I cry for this country. But, I continue to hope that the cloud will lift. Perhaps Saddam will yet turn tail and run. Perhaps reason will somehow still prevail. I along with millions of Americans will pray for the safety of our troops, for the innocent civilians in Iraq, and for the security of our homeland.
>
> May God continue to bless the United States of America in the troubled days ahead, and may we somehow recapture the vision which for the present eludes us.

If this kind of passion for truth and the courage to take a clear stand can be sustained by even a few of the senior leaders of an invading country – then a big peace movement can be mobilised which can stop any invasion.

Conquest and Peace
Challenges Ahead

IT IS NOT a mere coincidence that some of the worst events generally also bring with them a ray of hope.

When the forces of injustice become very aggressive, generally we find several groups of people, no matter how small and powerless, raising their voice to resist this injustice and protect the likely victims. It is these small groups of people who have always kept alive the faith and hope in humanity even in the worst of times and places.

These forces of justice and hope again asserted themselves as the Bush-Blair alliance started moving relentlessly towards an invasion of Iraq in 2003. On 15 Feruary, the city of London saw the biggest ever protest demonstration in its history, with a likely participation of around 1.4 million people. Huge protests were also seen in the cities of two important countries whose governments were supporting this invasion – Italy and Spain.

Despite the police creating many problems, US cities like New York also saw some very impressive protests. On a single day, 15 February, perhaps over 15 million people protested all over the world. On the whole, perhaps 10 times this number would've taken part in one protest demonstration or the other before and during the days of the Iraq invasion.

True, the protests did not stop the invasion.

Equally true, this cruel, ruthless invasion would've been perhaps even more cruel and ruthless if the protests had not taken place.

In these difficult and critical times, any effort for peace, no matter how small and inconsequential, is invaluable. Yesterday it was Iraq, today it is Iran. In the middle of many disappointments, let's never forget the value of our small efforts and those of our colleagues.

At the same time, there is no doubt at all about the need to make the peace movement stronger, more broad-based and effective. It is important to retain its continuity – in keeping with ever present threats – instead of sudden peaks and slumps. This can happen if the peace movement is alert not only to the calls for war, but also to the root causes of war and aggression.

Wendell Berry, the farmer-poet, philosopher and conservationist from Kentucky, USA wrote at that time of the Iraq war, in a paper titled, The Failure of War:

> For about half of the last century we worried about world conquest by international communism. Now with less worry (so far) we are witnessing world conquest by international capitalism. Though its political means are milder (so far) (note – this was written before the Iraq invasion) than those of communism, this newly internationalised capitalism may prove even more

> destructive of human cultures and communities, of freedom, and of nature.
>
> Its tendency is just as much towards total dominance and control. Confronting this conquest, ratified and licensed by the new international trade agreements, no place and no community in the world may consider itself safe from some form of plunder. More and more people all over the world are recognising that this is so, and they are saying that world conquest of any kind is wrong, period. (*Yes! magazine* and *Resurgence,* Nov-*Dec 2002.)*

Berry goes on to emphasise:

> Our right to live, to be free, and to be at peace is not guaranteed by any act of violence. It can be guaranteed only by our willingness that all other persons should live, be free, and be at peace – and by our willingness to use or give our own lives to make that possible.

To summarise, Berry has made three important points which the peace movement will do well to ponder.

(1) The increasingly aggressive trends in world economy are tied up inseparably with massive plunder of others resources, hence with conquest and war.

(2) More and more people in the world are recognising this reality, hence protesting.

(3) This protest, in order to be able to create a better world, should take the form of nonviolent, peaceful resistance – a task so important that, ideally, we should devote our life to this.

Economic Crisis and Social Distress

At a practical level, there is a clear need in conquerer societies to spread greater consciousness that the aggressiveness of their economic and political conquests, their quest for domination is self-destructive.

Once aggression, conquest, domination, plunder, deceit are accepted for amassing wealth from other regions and their people, then it is only a matter of time before such attitudes become acceptable and commonplace in the conduct of domestic and personal affairs. The great economic inequalities, social disruption and moral decay of most conqueror societies, from the ancient Romans to the present day USA, can be traced to a large extent to this factor – the self-destructive nature of conquest and domination.

If – at many practical levels – it can be shown that the acute social distress and economic crises within conquerer societies are closely linked to the their aggressive outside policies, then it will be easier to mobilise many more people against the policies of conquest and domination. *When more and more people – for example small farmers and workers in societies dominted by big corporations – begin to see clearly how the policies of domination that ravage the people of other countries also ravage them, then the peace movement's prospects of peacefully transforming these conquerer societies will improve significantly. The women's movement and environmental protection movement – two very important voices for justice – have close linkages with the peace movement which need to be explained more clearly.*

Among the societies where the tendencies for terrorist violence and counter- violence have spread, the peace movement must, of course, spread the message that such violence cannot achieve any worthwhile goals. However, this message will be effective only if alternative, peaceful platforms of protest can be created which are effective and which are supported by peace-loving people in conquerer societies as well.

The international anti-terrorism effort should be supported by the peace movement, but at the same time, also monitored carefully to ensure that it is not used as a pretext to serve the aims of conquest and domination (this is increasingly the case at present).

Within conquerer societies many people and professions are made to serve the goals of conquest and domination without realising the extent to which they have participated in plunder, murder and environmental destruction

These connections should be made clearer to more and more professional groups and their assistance sought in spreading awareness of professional ethics. Before the invasion of Iraq, the media was used very destructively in the USA to establish a linkage between 9/11 terrorist attacks and Iraq.

As Noam Chomsky has explained: before September 2002 only 3% of US people accepted this linkage. But a US government-supported media campaign created a situation in which within a few months as many as 50% of people accepted this linkage. The acceptance of this linkage contributed in a big way to the public accepting this invasion. Hence, wittingly or unwittingly, media persons played a very crucial role in the invasion of Iraq and many leading media persons have to accept a share of the blame for all the cruelty that was unleashed on the people of Iraq.

Awareness of such professional responsibility should increase and a lot of effort has to be made to ensure that there is adequate resistance within the professions and by professional ethics organisations to prevent such criminal misuse in future.

Peaceful Resistance to War
This can be a Reality

> Yet at the end of a century in which we have fought two wars to end war and several more to prevent war and preserve peace, and in which scientific and technological progress have made war ever more terrible and less controllable, we still, by policy, give no consideration to non-violent means of national defence. We do indeed make much of diplomacy and diplomatic relations, but by diplomacy we mean invariably ultimatums for peace backed by the threat of war.
> To achieve peace by peaceable means is not yet our goal. We cling to the hopeless paradox of making peace by making war.
>
> —Wendell Berry

IN THESE DAYS of the most cruel invasions and equally ruthless, thoughtless terrorist attacks, one doesn't hear much about the potential of peaceful, non-violent resistance by people. But it is important to keep alive these possibilities lest people forget entirely the path shown by Mahatma Gandhi and Martin Luther King.

Before the invasion of Iraq, I had issued an appeal, highlighting some possibilities of peaceful resistance to prevent this invasion. This is reproduced here for the consideration of peace-loving people at the time of the possibility of any such invasion anywhere in the world in future.

Prevent Invasion by Organising All-religion Prayers

Appeal made before the Iraq invasion

One last effort can perhaps still be made by the peace activists after securing the cooperation of Iraq government and perhaps some neighbouring countries as well. This effort should involve about one hundred thousand people (or more if possible) from Iraq, neighbouring countries as well as volunteers from Western countries and other parts of the world.

These entirely unarmed people will gather every day and night at about ten strategic locations in Baghdad (can be anywhere in the area under threat) and perhaps some other important parts of Iraq (or any other country threatened). These people from all parts and all religions of the world will continuouslypraye aloud and sing songs of peace. They will be entirely unarmed. They will not shout any hostile slogans. An atmosphere of complete peace and serenity will prevail at these gatherings.

It is hoped that these huge gatherings for all-religion prayers will be able to prevent the invasion. The most eminent peace-loving people from all walks of life will be called from time to time to bless these gatherings. At least one such eminent person will come every day. Those who are able to stay for a few days will remain here. Various peace-loving organisations from various parts of the world will be requested to send special delegations.

All of these people from various religions will together continuously pray for peace and harmony in the entire world. These prayers will continue till a firm decision is taken to cancel the invasion. While many peace-loving people will participate directly in these gatherings, others will take responsibilities such as arranging transport and food and ensuring that this message of peace reaches more and more people.

LINKING GLOBAL AND LOCAL

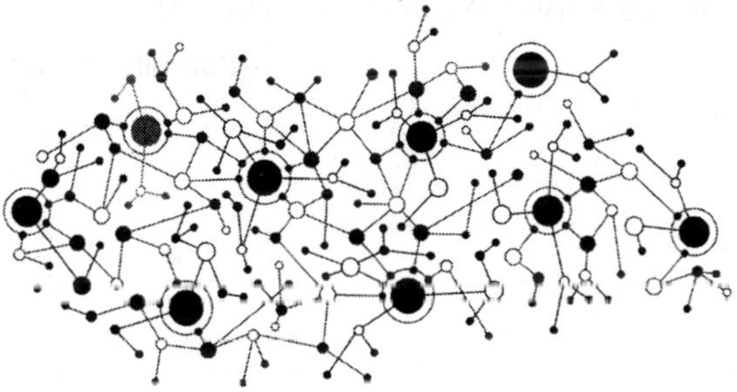

If humanity is to progress, Gandhi is inescapable. He lived, thought and acted, inspired by the vision of humanity evolving toward a life of peace and harmony. We may ignore him at our own risk.

—Martin Luther King Jr.

Gandhi is not only for India a hero of national history, whose legendary memory will be enshrined in the millennial approach. Gandhi has renewed for all the people of the West, the message of their Christ, forgotten or betrayed. …I have seen here, in Switzerland, the pious love that he (Gandhi) inspired in humble peasants of the countryside and the mountains.

—Romain Rolland

Survival Problems

Gandhi's Message for Resolving Crisis

THERE IS GROWING realisation that our world is increasingly faced with a survival crisis in the sense that the very life-nurturing conditions of our planet which made such amazing biodiversity possible on earth are threatened.

Time-bound solutions have to be found within the next two decades to prevent several of these threats from crossing the much-feared tipping points beyond which these may slip out of human control. As finding time-bound solutions for these threats becomes the biggest responsibility of humanity, relevance of Gandhiji's core ideas has increased even more than before as there can be extremely useful for providing the basic value system which is most conducive for resolving the survival crisis.

The Survival Crisis

In 1945, after the use of the first two atomic bombs it was realised, perhaps for the first time, that human action can be so destructive as to threaten the life-nurturing conditions of our planet. This feeling accentuated with the subsequent accumulation of nuclear weapons and availability of more information about the potential destruction chemical and biological weapons can also cause.

In the 1970s, information started becoming available that human-made environmental changes can also be highly threatening for the basic life-nurturing conditions of our planet. This understanding become more definitive from 1990s onwards and resulted in the publication of IPCC Reports. Apart from climate change, about 10 survival threatening environmental issues can be mentioned in this context. The concept of 'tipping point' tells us that these problems have to be resolved within a very limited time-frame or else the situation may go out of humanity's hands.

The United Nations Environment Programme issues periodic reports on the state of the world's environment, recent trends and future prospects. The Global Environment Outlook 5 has presented 'undeniable evidence that the world is speeding down an unsustainable path'. This report has voiced a clear warning that urgent changes are needed 'to avoid exceeding critical thresholds beyond which abrupt and generally irreversible changes to the life support functions of the planet could occur'.

This issue of critical significance has been taken up in greater detail in the work of scientists at the Stockholm Resilience Centre (SRC). Johan Rockstrom, director of SRC says,'The human pressure on the Earth System has reached a scale where abrupt global environmental change can no longer be excluded. To continue to live and operate safely, humanity has to stay away from critical "hard-wired" thresholds in the Earth's environment, and respect the nature of the planet's climatic, geophysical, atmospheric and ecological processes'.

The scientists at SRC first identified the Earth System processes and potential biophysical thresholds, which, if crossed, could generate unacceptable environmental change for humanity. They then proposed the boundaries that should be respected in order to reduce the risk of crossing these thresholds. The nine boundaries identified were: climate change, stratospheric ozone, land use change, freshwater use, biological diversity, ocean acidification, nitrogen and phosphorus inputs to the biosphere and oceans, aerosol loading and chemical pollution. The study suggests that 'three of these boundaries (climate change, biological diversity and nitrogen input to the biosphere) may already have been transgressed'. In addition, it emphasises that 'the boundaries are strongly connected – crossing one boundary may seriously threaten the ability to stay within safe levels of the others'.

The period 1990-2018 is a period of lost 28 years in the sense that the seriousness of the survival crisis was well-realised but nothing very effective was done to check this in these many recent years. In fact, various aspects of the survival crisis continued to become more serious during this period.

The last one or two years have been particularly bad from this most crucial perspective as the most powerful and resourceful country (the USA) has moved away from the solutions more than ever before. Due to a complex of reasons, the threat of actual use of weapons of mass destruction has increased, while robot weapons (AI or autonomous weapons) are also fast emerging as a major threat.

While these various issues are generally discussed in their individual context, at a practical level, the world will increasingly face these accentuating problems together at the same time. Hence, there needs to be a much greater sense of urgency regarding these most serious problems at the world level. If the present world leadership is not according adequate and appropriate importance to these issues (as is certainly the situation now), future hope of resolving these issues will be cantered mainly on people's efforts and actions.

Value-System Needed to Resolve Crisis

While technological solutions are being proposed for these various problems and international conferences are organised regularly to provide a detailed framework for this, it is increasingly realised that we are lagging much behind in the tasks of highest urgency.

As the United National Secretary-General's High-level Planel on Global Sustainability stated, 'Despite the adoption of the United Nations Framework Convention on Climate Change and its Kyoto Protocol, annual global carbon dioxide emissions from fuel combustion grew by about 38% between 1990 and 2009, with the rate of growth faster after 2000 than in the 1990s'.

The agreements on curbing accumulation and possibility of use of weapons of mass destruction appear to be weaker today than a few decades ago.

One main reason of this failure is that the prevailing value systems in the present-day world are not conducive to resolving the survival crisis. If the existing value systems do not become favourable to resolving the crisis in time, the search for immediate technical-fix solutions in the form of highly dubious huge geo-engineering changes may be pushed by authoritarian regimes with their own narrow agenda. Our effort with the greatest sense of urgency should be to find solutions for the many-sided survival crisis within the framework of democracy, peace and justice.

To give just one example, if the desire of ever-increasing consumption, including wasteful and harmful consumption continues to remain unchecked, then despite technological innovations in the area of replacing fossil fuels success in reducing overall GHG emissions can elude us, or else limited success in this can be accompanied by increasing problems in other areas.

On the other hand, if such technological solutions are accompanied by willingness for overall very significant reduction in consumerism, luxury consumption and wasteful consumption then along with very significant reduction in GHG emissions several other aspects of environment protection can be achieved at the same time, along with other significant benefits such as improved health and reduced mortality rates.

In particular, the biggest scope may be for reducing production and use of small and big arms, including weapons of mass destruction. If the overall value systems in world strongly favour peace and are against all forms of violence, then it may be possible to reduce all arms and ammunition production and use by about 80 to 90 per cent.

There is clearly a strong need for spreading values of peace and non-violence directly. In addition, however, the efforts for reducing consumerism and greed can also create conducive conditions for peace and non-violence. As greed decreases, so does the inclination at all levels to snatch the share of others.

This leads to a much lesser inclination for violence and provides much more favourable conditions for spread of values of peace and avoiding use of force.

Widespread conditions of peace were always needed, but these will be increasingly needed all over the world to resolve survival issues and to secure international cooperation for this.

In brief, then, there is clearly a need for efforts and campaigns which create in society a base of values which are favourable for resolving the survival crisis. Values of frugality and simplicity and curbing endless consumerism on the one hand and values of peace and non-violence are very important and essential components of this value system that we need. These two components are also closely inter-related.

Ideas of Mahatma Gandhi

Both of these components of value systems were emphasised repeatedly by Mahatma Gandhi. These are so closely identified with him and rightly so, that anyone being emphatic about these two aspects is often referred to as 'Gandhian' in India.

Mahatma Gandhi was a most remarkable man of many-sided accomplishments but is best known for his life-long resistance to injustice using an approach of complete and uncompromising non-violence. This inspired numerous other struggles including many small and big social and environmental movements in post-independence India. Gandhiji's example also inspired many important non-violent struggles outside India, including those led by Martin Luther King and Nelson Mandela.

This legacy of Mahatma Gandhi will be extremely useful in future for people's movements to influence governments and institutions to make the changes which are necessary for resolving the survival issues in time. The kind of moral and ethical force which Mahatma Gandhi could bring to these non-violent movements will be increasingly needed by these movements.

The overall emphasis on non-violence in all aspects of life and as a way of life which Gandhiji emphasised can be very valuable for creating the base of a peaceful world. Widespread conditions of peace are of course highly desirable in themselves, but in addition will be increasingly needed to meet the challenges of various survival issues.

In a world inspired by the principles of non-violence and peace at all levels, the chances of reducing arms and ammunition to a significant extent and for the success of disarmament agreements at various levels will be much higher.

Gandhiji placed a strong emphasis on voluntary limitation of consumption by people by basing it on need rather than greed. This reduces greatly the chances of conflict and improving the prospects of peace because there is less likelihood of snatching the resources of others. Gandhiji has become a symbol of simple and frugal life and yet could get the long-term following of millions. Given the trends of present-day populist politics, this is a lesson that things can be done differently.

At a time of increasing trends of urbanisation and centralisation, Gandhiji placed increasing emphasis on decentralised rural pattern of life. He emphasised the creation of self-reliant rural communities. These communities should be very creative in terms of making the best possible use of local resources and skills for meeting as many needs of daily life as possible. The carbon costs of transporting food and daily need goods would be very low. It is in these patterns of life that we've the most scope for community action for reducing greenhouse gas emissions in many ways. He also spoke against tendencies of over-mechanisation, which often involve higher GHG emissions, and a greater scope for human creativity and skills.

Gandhiji's experiments with community living may be increasingly relevant in new contexts which bring out how basic needs of all people can be met in healthy and harmonious ways on the basis of sustainable livelihoods , reducing the chances of all conflicts while keeping GHG emissions at extremely low levels.

The great need of today is for integrating the agenda of justice with agenda of environment protection and here again Gandhi's ideas are very helpful as he placed the needs of the poorest people above everything else. His concept to trusteeship is also significant in this context.

As the world grapples with the increasingly difficult task of finding time-bound solutions to various aspects of the survival crisis in the near future it will no doubt look out with an increasing sense of urgency for inspiring figures in human history whose ideas and work can be useful and inspiring for meeting this challenge. In this context, there is no doubt that there will be an increasing need and scope to learn from the work and ideas of Mahatma Gandhi and their relevance for the 21st century will continue to grow more and more.

Hence there is a clear need to examine various concepts, ideas and work-experiences of Gandhiji in the more contemporary context of learning and inspiration for resolving most urgent issues, particularly those relating to various aspects of the survival crisis.

This paper was first written for Ekta Parishad and the Jai Jagat Campaign.

Non-Violence
A Way of Life

IN A WORLD increasingly plagued by violence and strife, Mahatma Gandhi's message of non-violence remains highly relevant today, 71 years after the assassination of this man of peace.

Gandhi's work and writings covered a very wide range of activities, but perhaps what our troubled world needs to learn most from Gandhi is his overwhelming emphasis on non-violence to resolve any conflict or to confront any injustice. More than that, he emphasised non-violence as an entire way of life, emphasising its importance in our daily life (even relationships within family) as much as in huge movements for justice and freedom. Non-violence (ahimsa) for Gandhi is all-encompassing, bringing in its range not only all human beings but also all forms of life. Non-violence in action is considered far from adequate – it should extend to our innermost thoughts.

Gandhi's concept of non-violence is so pervasive that it will not only help the cause of world peace – rightly practised, it will help to significantly bring down the incidence of domestic violence and mental illness as well. The movement for animal rights or compassion to animals will also benefit from a wider spread of Gandhi's ideas.

About the role of non-violence in his own life, Gandhi wrote:

> I have been practising with scientific precision non-violence and its possibilities for an unbroken period of over fifty years. I have applied it in every walk of life, domestic, institutional, economic and political. I know of no single case in which it has failed. Where it has seemed sometimes to have failed, I have ascribed it to my imperfections (*Harijan*, 6-7-40, pp. 185-86).

Gandhi had big hopes from harnessing the tremendous force of non-violence. He wrote (*Harijan* 10.12.38 pp. 377), 'More powerful than all the armaments, non-violence is a unique force that has come into the world'. Elsewhere he said:

> We are constantly being astonished these days at the amazing discoveries in the field of violence. But I maintain that far more undreamt of and seemingly impossible discoveries will be made in the field of non-violence (Harijan, 25.8-40, pp. 260).

The path of non-violence is noble, but it is not easy. As Gandhi says (*Harijan* 14.5.38), 'If the method of violence takes plenty of training, the method of non-violence takes even more training and that training is much more difficult than the training for violence'.

> A votary of Ahimsa has to be incorruptible, fair and square in his dealings, truthful, straightforward and utterly selfless. He must have also true humility (*Harijan*, 20.5.39, pp.133).

> The very first step in non-violence is that we cultivate in our daily life, as between ourselves, truthfulness, humility, tolerance, loving kindness. Honesty, they say in English, is the best policy. But in terms of non-violence, it is not mere policy. Policies may and do change. Non-violence is an unchangeable creed. It has to be pursued in face of violence raging around you (*Harijan*, 2.4.38, pp. 65).

Non-violence, to be a potent force, must begin with the mind (*Young India*, 2-4-31, pp. 58).

Unless there is a hearty cooperation of the mind, the mere outward observance will be simply a mask, harmful both to the man himself and to others. The perfect state is reached only when mind and body and speech are in proper coordination (*Young India*, 1.10.31, pp. 287).

> The alphabet of Ahimsa is best learnt in the domestic school, and I can say from experience that, if we secure success there, we are sure to do so everywhere else (*Harijan* 21.7.40, pp. 214).
>
> If one does not practice non-violence in one's personal relations with others and hopes to use it in bigger affairs, one is vastly mistaken. Non-violence, like charity, must begin at home (*Harijan*, 28-1-39, pp. 441).

Gandhi captured the spirit of non-violent struggle in the following words, 'I seek entirely to blunt the edge of the tyrant's sword, not by putting up against it a sharper edged weapon, but by disappointing his expectation that I would be offering physical resistance. The resistance of the soul that I should offer instead would elude him. It would at first dazzle him and at last compel recognition from him, which recognition would not humiliate him but would uplift him'.

For Gandhi, religion was clearly a place for peace and there was absolutely no room in his thinking for religious differences turning violent. He wrote, 'I believe in the fundamental truth of all great religions of the world. I believe that they are all God given, and I believe that they were necessary for the people to whom these religions were revealed. And I believe that, if only we could all of us read the scriptures of different faiths from the standpoint of the followers of those faiths, we should find that they were at bottom all one and were all helpful to one another'(*Harijan* 16.2.34 pp.5-6).

Gandhi saw clearly that the aggressiveness of the great powers is rooted in greed. He said world peace, 'is clearly impossible without the great powers of the earth renouncing their imperialistic designs. This again seems impossible without these great nations ceasing to believe in soul-destroying competition and to desire to multiply wants and therefore increase their material possessions' (*Harijan* 16.5.36 pp.109).

The ideal he urged to his own countrymen was. 'we will exploit none just as we will allow none to exploit us' (*Young India* 16.4.31 pp. 9). He added, 'I should reject that patriotism which sought to mount upon the distress or the exploitation of other nationalities'.

Vision of World Federation

In fact, Gandhi was an internationalist much ahead of his times.

He wrote, 'The better mind of the world desires today not absolutely independent States warring one against another, but a federation of friendly inter-dependent States' (*Young India*, 26-12-24, pp. 425).

'The structure of a world federation can be raised only on foundation of non-violence, and violence will have to be totally given up in world affairs (Gandhiji's Correspondence with the Goverenment - 1942-44, 1957, pp. 143).

'Federation is undoubtely a greater and nobler end for free nations. It is a greater and nobler end for them to strive to promote Federation than be self-centred, seeking only to preserve their own freedom' (*Harijan*, 9-8-42, pp. 265).

Socialist Society based on Non-violence

Speaking optimistically of his vision of a future world, he wrote:

> The world of tomorrow will be, must be, a society based on non-violence. That is the first law: out of it all other blessings will flow.
>
> An individual can adopt the way of life of the future – the non-violent way – without having to wait for others to do so. And if an individual can do it, cannot whole groups of individuals? Whole nations? Men often hesitate to make a beginning, because they feel that the objective cannot be achieved in its entirety. This attitude of mind is precisely our greatest obstacle to progress an obstacle that each man, if he only wills it, can clear away.
>
> Equal distribution – the second great law of tomorrow's world as I see it – grows out of non-violence. It implies not that the world's goods shall be arbitrarily divided up, but that each man shall have the wherewithal to supply his natural needs, no more (The Mind of Mahatma Gandhi, pp.458-60).
>
> Only truthful, non-violent and pure- hearted socialists will be able to establish a socialistic society in India and the world. To my knowledge, there is no country in the world which is purely socialistic. Without the means described above, the existence of such a society is impossible (*Harijan*, 13.7.47, pp.232).

Harijan and *Young India* were journals with which Gandhi was closely associated.

Jai Jagat
Universe as Family

IN RECENT DECADES our world has been increasingly living in the shadow of a huge contradiction. On the one hand the capability to almost eliminate poverty, hunger and deprivation appears to be growing. On the other hand, a number of very serious problems threaten to tear apart even the progress already achieved.

This contradiction appears to be getting even more acute in the early phase of the 21st century. On the one hand, the agenda of Sustainable Development Goals (SDGs) points to a path of rapidly diminishing distress and deprivation. Yet, on the other hand, increasing evidence continues to emerge that the very life-nurturing conditions of our planet are badly threatened. One aspect of this relates to climate change and a number of other serious environmental threats. The other aspect relates to accumulation of weapons of mass destruction, including the fast emerging robot, AI or autonomous weapons.

It is a matter of deep concern that on some of the most pressing issues which could potentially be most destructive – the progress has been far from satisfactory, and in some contexts the progress made earlier appears to have slowed down, stopped or even regressed in more recent times. We are far from meeting greenhouse gas reduction targets. Some of the agreements for control of weapons of mass destruction are not being renewed.

At this critical juncture humanity needs more than ever before very significant new initiatives which can check these highly destructive trends while at the same time channelising human capabilities towards the elimination of poverty and deprivation, as expressed in SDGs or in other ways.

One very important initiative has to be in the form of changing the value systems which guide and motivate human action. The prevailing dominant value systems have brought us to the destruction we face today and to try to check these destructive trends while continuing to retain the same value systems will not take us very far. Some of the basic changes that are needed are listed below.

Commitment to 'Universe as Family'

Our world needs to be guided and motivated increasingly by the concept of universal family, meaning that the commitment is to the welfare of the entire world without any discrimination. Earlier commitments have been to the nation-state, or at other levels have been guided by religion, region or other commitments. However, a much wider commitment is increasingly needed now – commitment to the entire universe without discriminating in any way among the various people of our world. This is needed because without the spread of the basic value of 'entire universe as family' it does not appear likely that it will be possible to solve the most pressing problems.

As long as most people in the world are guided by narrower commitments, the negotiations and efforts of solving the most serious problems remain very restricted and within this narrow arena the chances of real success are very less, as has been proven time and again in recent times. On the other hand, when more and more people are guided by the concept of 'universe as family', then many very exciting possibilities of resolving serious world problems are opened up.

Commitment to Peace as a Way of Life

This is a commitment which has been often preached but commonly neglected in practice, as is evident from the enormous toll of violence in every- day life in family, in neighbourhood, in school and college, in workplace and on the street. Violence (including verbal and emotional violence and also including violence to deprive others of their rights) is perhaps the biggest cause of distress in the world, including deep emotional distress which has long-term impact. Gender-based violence of various kinds is by itself a huge tragedy, causing physical harm and emotional distress on a massive scale almost all over the world.

At a wider level, tendencies of violence are reflected in wars and civil wars, arms race, accumulation and threat of weapons of mass destruction. While these have already caused massive distress, the destructive capacity of war, civil war and terrorist actions have been greatly enhanced with the emergence of even more destructive weapons. The most destructive weapons available now can destroy most life on earth many times over. Due to a complex of factors, the likelihood of actual use of weapons of mass destruction has been increasing in recent times.

A deeper commitment to peace and non-violence is thus most urgently needed. This is a must for resolving the survival crisis. A huge reduction in the possibility of war, accompanied by reduction in preparation for war and arms production and deployment, will also have very important spin-offs for environment protection and meeting real needs of people.

If a deep and firm commitment to peace and non-violence spreads widely, among the people of world, this can make the most important contribution to reducing distress in everyday life. Further this commitment of a very large number of people to peace, if firmly established, can be expected ultimately to create conducive conditions for a big reduction in the possibility of war and arms race.

Cooperation in Place of Dominance

The more prevalent tendency among human beings has been to try to dominate each other, at various levels and in various ways. This should be replaced by a spirit of cooperation and understanding each other. Efforts made in this direction will have many creative manifestations, not the least at the level of gender relations, and contribute to much joy and happiness (as well as reduce possibilities of conflict and tension) in school and college, family, neighbourhood and workplace.

The impact of this will be reflected in increasing cooperation among communities, provinces, regions and nations. Chances of conflict at all these levels will decrease.

Commitment to Protection of All Forms of Life

Earth is not just for human beings, it is for all forms of life which live on this planet. As the most capable species on earth, human beings should have a protective attitude towards various forms of life. There should be a commitment also to protect all habitats and the wider environmental conditions without which various life forms would be threatened. While humans have a reasonable right to more specially protect their own interests, these should be defined within those parameters which are protective for all forms of life.

Commitment to Earth and its Life-nurturing Conditions

Earth has special conditions – mix of various gases, water, temperature limits etc. – which are essential for the survival and flourishing of a vast diversity of life species. These life-nurturing conditions are not guaranteed by a super-natural force, but have to be maintained and protected by human beings. During the last few decades, human beings have shown an unprecedented capacity to disrupt and destroy these life-nurturing conditions and this can be most destructive for future generations (and to some extent is already proving destructive for the present generation). Hence, a commitment to protecting Earth and its life-nurturing conditions is more needed today than ever before.

Commitment to Justice and Democracy

While the world is faced with very urgent issues of survival which demand priority attention, it should be clear that all such efforts should be within the framework of justice and democracy.

The urgent need to reduce greenhouse gas emissions should be tied up with a plan to ensure basic needs of all people, a path which is along the lines of SDG thinking but in practice has been increasingly bypassed as justice concerns have been receiving lower attention than before in climate change negotiations. With carbon constraints and other environmental constraints added and the need to limit fossil fuels and also all material goods beyond a point, the need for justice-based equality and equality-based justice has increased more than ever before.

As greenhouse gas emission cuts and related environmental norms are stepped up, priority must be given to the production of basic needs of the poor and not luxuries of the rich. Wasteful products and products which harm health and environment should be cut down significantly. By linking justice issues, possibilities of getting support of a much larger number of people for resolving issues of great urgency increase significantly.

Along with justice, a strong commitment to democracy is needed. When issues of great urgency have to be resolved quickly, there is a temptation to neglect democratic norms, but it is only by making efforts to take this issue to a very large number of people in a big yet meaningful way that a strong and sustainable base for effective solutions can be created.

Big decisions will have to be taken by critically-placed institutions, but only democratic actions can ensure that they are the most effective decisions taken under the watchful eyes of people. In the process, pressures are also built up for long overdue reforms of institutional structures (including crucial international institutions), so that either their capacity to resolve the most urgent issues is enhanced, or else new institutions capable of resolving the emerging most urgent issues are created in democratic ways.

Conducive Value System to Bring Change

Six aspects of changing value systems visualised above can together contribute to creating a social base which will help to bring and sustain public support for urgently-needed changes to resolve various pressing survival issues. At the same time, these value systems will help tremendously and in very creative ways to reduce distress in daily life of millions and millions of people, and this success will provide the basis for sustaining these value systems. Once these value systems reach a certain level, these have the capacity then to spread on their strength, on their capacity to reduce distress in daily life and also make it more meaningful, creative and satisfying in a deeper, different way.

In the absence of such social values as a base of public-support and action, numerous technological innovations, international conferences and financial commitments, no matter how well-intentioned and even essential, are unlikely by themselves to succeed in bringing and sustaining the changes necessary to resolve the most urgent issues.

Jai Jagat as a Movement to Spread Gandhian Values

Several people, organisations and movements all over the world are working in their own small but significant ways to strengthen various aspects of these value systems.

These efforts, while highly valued and desirable (although often somewhat isolated), have been important yet far from adequate to bring and sustain significant changes in a well-integrated way. The various aspects of value-systems emphasised are closely related to each other and are most effective as a well-integrated system which brings together the concerns of peace, environment protection, justice and democracy and channelises them effectively to resolve most urgent issues.

The concept of Jai Jagat seeks to bring and sustain a convergence of these closely related values to create a wide social base for resolving the most urgent issues of the world within a framework of justice and democracy. The *Jai Jagat Movement*, based on this concept, seeks to mobilise people across all narrow divides and boundaries so that in the spirit of 'universe as a family' people can be united in resolving the most urgent survival issues in just and non-violent ways which also help significantly to reduce distress and deprivation in daily life.

Jai Jagat seeks to make this mobilisation in the entire world on the basis of equality of all and absolutely no discrimination on any basis, whether gender, colour, religion, caste, ethnicity or nationality. If all over the world there are more and more people with a firm faith in these value systems, then significant social mobilisations necessary for resolving the most urgent issues in peaceful and just ways will not only emerge but will also be able to sustain their efforts.

Such social mobilisations are most necessary for these changes, for while the formal decisions will be announced at the institutional level it is social mobilisations who will influence these decisions and also help to ensure that the announced changes are implemented in the right spirit and with the necessary urgency. As over-dependence on isolated institutional action has proved very costly in the past, the importance of such social mobilisations beyond national and other narrow boundaries is evident. At the same time, it is not possible to sustain such social mobilisations without a strong base of conducive value systems.

This is where the Jai Jagat movement (which seeks to strengthen these value systems all over the world, keeping in centre the concept of 'universe as family') can fulfil a very important role at this most critical juncture of earth's history and fill an important gap in the overall efforts for resolving most urgent issues.

Inspired by Gandhi

Jai Jagat draws its inspiration from several personalities and movements in history, but clearly in terms of its inspiration sources there is a centrality of the life and ideas of Gandhiji as well as the various movements and mobilisations (including Sarvodaya) inspired by him.

Moreover, it is in Sarvodaya circles that the greeting and guideline of Jai Jagat has been most frequently used, to convey sincerely held beliefs of 'universe as family' 'well-being of all', 'peace to all' etc. Here – as an inspiration source – we must particularly include the name of Mahatma Gandhi's dear colleague Vinoba Bhave. His commitment to Jai Jagat, particularly 'universe as family' principle, is well-known, while he is even better known for his path-breaking work of bhoodan or gift of land.

A firm faith in peace and non-violence based on justice is more needed than ever before for resolving the most urgent issues. For Mahatma Gandhi, a firm commitment to peace and non-violence was the most basic principle. On the strength of this basic principle he, with his close colleagues, launched several struggles for justice which could mobilise tens of millions of people and which continued for over half a century.

His work continues to inspire tens of millions of people over 70 year after his death and is often used as an inspiration and learning source when very difficult situations and critical questions have to be resolved.

Even while fighting highly-oppressive colonial rule for three decades, Mahatma Gandhi never spread any hatred against oppressors and remained true to the concept of 'universe as family'. The member of the family who had turned rogue and was inflicting injustice on other members had to be confronted and opposed in peaceful ways for the sake of bringing justice, but nevertheless s/he was still accepted as a member of (the universal) family.

Gandhi went beyond contemporary debates of capitalism vs communism to focus on the more basic civilisational issues of avoiding and resisting greed, discrimination and hatred at various levels and the desirability of a simple, frugal, principled and truthful life.

He emphasised compassion which covered all forms of life. He emphasised that the poorest and the most deprived people must be the centre of our concern while deciding about priorities. He could foresee some of the most serious problems (including environmental ones) much earlier.

Unlike other political leaders, he did not promise the moon to his followers, but instead emphasised frugality and simplicity. He set up several community living (ashram) experiments where many ideas of simple, frugal, cooperative, non-discriminative, low resource-based, more livelihood protective, less costly and more replicable patterns of living could be tried out by like-minded people.

Hence his life, work and vision offer a particularly inspiring source for present-day efforts which seek to combine concerns of peace, environment protection, justice and democracy to find effective solutions for some of the most urgent issues which confront our world today. Of course, there is enough room for learning from many other sources of inspiration – persons as well as and movements.

Dialogue and Mobilisation

There is clearly need for dialogue of various kinds at various levels to spread the idea and concept of Jai Jagat. This well-informed dialogue should have the necessary research and documentation support with focus on carefully identified significant questions and issues. From this dialogue many new ideas and volunteers will emerge, various persons and organisation will take up certain responsibilities. Many persons and organisations are working with more or less similar viewpoints, but in somewhat isolated ways. They may be eager to join. Many may feel that a wider opportunity to take forward their strongly-held beliefs has become available. A decision may be taken to make a more concentrated effort in some areas, given the limitations of resources and time, to try out how far the Jai Jagat concept initially secures the involvement of people in various sections of society, what is their response, what is their level of support, how they can contribute, what are their objections (if any). These experiences and interactions should be carefully documented as learning experiences.

The dialogue is also linked to mobilisation, of course.

A significant aim is to secure the involvement of people who will devote all, most or some of their time to working with and for Jai Jagat. There should be adequate preparation to provide creative avenues to the people, particularly youth, who join, the aim being to make the best use of their overall commitment but also of their more specific or specialised skills. Organisations working along somewhat similar lines can certainly continue to retain their present identity while linking up with Jai Jagat for various initiatives and efforts.

There should be an emphasis of sustainability of efforts and longer-term commitments from the outset. Involvements should not end with any big event. A path of contributing on a continuing basis in daily life (not just in special events) should be clear to Jai Jagat volunteers and activists. Jai Jagat as a way of life should be emphasised.

Inter-action with Institutions

While social mobilisation is a key role of Jai Jagat, inter-actions with key institutions (including international institutions) and institutional decision- makers cannot be ignored. This is particularly so at a time of urgent survival issues with their tipping points crying out for time-bound solutions, formal decisions for which will ultimately have to be announced by institutional decision-makers.

Social mobilisation along Jai Jagat lines can pave the way for public action which pushes institutional structures towards badly-needed changes, both in terms of crucial decisions and their own internal reforms. Once institutional decisions have been taken, social mobilisation helps in their better and speedy implementation in the right spirit.

In particular, Jai Jagat mobilisations will be critical for overcoming the constraints posed by narrow considerations so that a 'universe as family' approach to the resolution of problems (which threaten life-nurturing conditions in the entire universe) is possible.

At present, institutional efforts to tackle life-threatening issues, such as climate change and weapons of mass destruction, are constrained by a narrow agenda which can be pursued within the present narrow framework. Several more obvious solutions are outside this framework. For example, greenhouse gas emissions can be reduced to a significant extent if arms and ammunition production can be reduced by 80 per cent or if the possibility of any war or civil war is drastically reduced.

A drastic reduction in many other harmful products is also possible. In the negotiation for reducing weapons of mass destruction, more ambitious (and highly desirable) goals like complete and total elimination of nuclear weapons and saying a complete 'no' to upcoming robot weapons can be possible. So, on the one hand, much more ambitious (and desirable) goals can be introduced in the negotiations, and, on the other hand, the ability of social mobilisation to influence these decisions is also increased, and at the same time there is internal reform of institutions to make them more responsive to more ambitious changes. This can bring much more hope for effective decisions concerning critical survival issues to be taken and implemented before it is too late.

Self-reforms

No less important is change in the life pattern of persons who are a part of this mobilisation. Are they only pushing for change or are they also reforming and changing their own lives and life-patterns in line with the principles of the change they want to bring? This will be a crucial question that the society will be considering while assessing and evaluating this mobilisation.

It is on this basis that the ethical strength of the mobilisation will be assessed, and it is on this basis that more and more people will be taking their decision to join this social mobilisation. What kind of relationships members of mobilisation establish among themselves, to what extent these relationships reflect a spirit of genuine cooperation and avoid dominance, to what extent various sections of the mobilisation rise above any narrow interests to advance the wider cause – all this will be important for assessing the sincerity of the social mobilisation and its members.

Self-reform is a very integral part of such a social mobilisation. The basis of self-reform is drawn from the wider principles of 'universe as family', peace and non-violence, environment protection, justice and democracy. Personal life should be as far as possible in keeping with these principles. Keeping the principles at centre, the actual life-pattern will, of course, differ for various regions, cultures, faith, climatic conditions, geographic and historical contexts.

Maintaining personal freedom and choice is very important in itself, as also are local and cultural contexts. So, self-reform certainly does not mean the quest for any uniform lifestyle. What it means is that people engage in a serious introspection and discussion on what should be a desirable pattern in keeping with the principles of peace, environment protection, justice, democracy and 'universe as family' and then strive to come as close to a life pattern based on these principles as is possible within their local and cultural context.

Strengthening Peace and Non-violence

The most important contribution of Jai Jagat has to be to the agenda of peace and non-violence, both because of its historic legacy and also because of the pressing need of these times.

While the agenda of peace and non-violence has always been important for the world, today it is essential for sheer survival. This is partly because of the growing destructiveness of weapons and their capacity to destroy all life. This is also because of the growing risks of proliferation and actual use of nuclear weapons due to a complexity of factors. In certain circumstances, even terrorists can kill tens of thousands of people using chemical or biological weapons or even crude nuclear weapons. Robot weapons bring a new range of problems including, in certain circumstances, the possibilities of some of these weapons even working outside human control.

Leaving aside the weapons of mass destruction, even the destructive potential of small arms has grown enormously despite efforts, including international treaties, to curb the proliferation of small arms.

There are also other wider reasons why an agenda of peace and non-violence is critical to resolving survival issues at this juncture of human history. Climate change and a range of other serious environmental problems have taken the form of a survival crisis. To resolve these, increasing and sustained international cooperation is needed without any disruptions. Very wide-ranging, concerted and continuing efforts are needed by all nations and communities.

Hence, these issues cannot be resolved in a world which is in a constant state of tensions, suspicions and rivalries, where there are several wars, civil wars and riots while the threat of several other conflicts looms over the horizon. This is in addition to the fact that wars, war preparations and arms production for wars are themselves very big pollutants.

The big environmental issues including climate change cannot be resolved in a world ridden with conflicts and threats of conflicts. We increasingly need conditions of continuing justice-based peace and stability in which continuing, concerted efforts can be made by all communities and nations to resolve survival issues with increasing cooperation among themselves in a spirit of 'universe as family'.

It is in this context that the social mobilisation of Jai Jagat and related efforts, with their historic legacy and linkages with Gandhian thinking and struggles, can make a very important contribution to strengthening a strong peace movement in our troubled world. When threat of war and invasion emerges, millions can come out to protest this. But such one-time protests, the event-based peace movements are not adequate. A peace movement which involves millions with continuity on a sustainable basis is needed.

Ordinary people understand the threat of nuclear weapons, but they legitimately ask – what can we do about this? They must get avenues where they can link up with the efforts for big changes.

To involve millions, the peace movement must spread non-violence as a way of life. If it can succeed in reducing the massive distress caused by violence in daily life (in school and college, street and work place, family and neighbourhood), millions will benefit and will be happy to join the peace movement as a movement which reduces distress and improves social relationships. At this stage, as the peace movement grows, it acquires the capacity to influence the big decisions on war and arms race and weapons of mass destruction. Emphasis on peace among people practising different faiths also helps in this process. Hence all the daily concerns and big macro concerns of peace can be incorporated into a broad-based, all-inclusive peace movement which has continuity and sustainability.

At the same time, the peace movement should build close linkages with other movements. With its wide concerns, Jai Jagat is in a good situation to facilitate such integrations, vertically and horizontally, to give the peace movement a very important presence.

Linkages of Peace and Other Movements

By building close linkages with other movements, the peace movement is further strengthened and so are other movements.

As emphasised earlier, the peace movement has close links to the environmental movement. To resolve big environmental issues, we need conditions of peace and stability. Wars and arms race are in themselves very big polluters and greenhouse gas emitters. On the other hand, to establish sustainable peace, urgent environmental issues need to be resolved as otherwise water crisis, weather extremes, breakdown of livelihood systems and sharing of increasingly scarce resources can increase old conflicts and trigger new ones. Hence, at many levels, the peace movement and environmental movement meet and Jai Jagat can contribute to enhancing their cooperation.

The movement for equal rights of women is very important in itself, but its increasing linkages with peace movement will benefit both. The peace movement, by spreading non-violence as a way of life, can help to create conditions in which domestic violence and sexual violence can decrease significantly. It can help to reduce all violent conflicts in which women and children suffer the most. On the other hand, women can be a very big source of strength for the peace movement.

The linkage with movements for justice are also obvious. Conditions of peace and stability are most conducive for meeting the basic needs of all people. Reduction of inequalities and injustice provide better conditions for achieving peace and stability.

With its wider concerns and mobilisation, Jai Jagat can play an important role in strengthening linkages among the various social movements.

Sustainable Development Goals

SDGs are admirable aims which, it is hoped, will help to prioritise the reduction of deprivation and environmental ruin. But in a business-as-usual scenario, there are increasing doubts whether these can be achieved in a world ridden with nothing less than a survival crisis.

However, if very important initiatives for establishing conditions of peace and stability with justice are taken, if there is a resurgence of social movements for justice and environment protection, if there is important institutional reform, if timely action is taken to resolve survival issues then conditions of realising (and in some contexts even surpassing) SDGs can emerge.

Jai Jagat can make an important contribution in this.

Significance and Sustainability

Clearly, the Jai Jagat movement has a very important role in a world deeply troubled with a multi-dimensional crisis. With its core values of 'universe as family' and peace and non-violence as a way of life, Jai Jagat can make a very important contribution in resolving several of the most urgent problems facing the world today.

However, for this aim to be realised, it is necessary to create a wide social base of Jai Jagat with continuing creative activities of high social relevance. Troubled people of the world, particularly youths, are looking for such spaces where they can contribute their creativity and energy for protecting the planet and creating a better world based on peace, justice and democracy.

The 'universe as family' concept was needed in all historic times. Commitment to peace and non-violence as a way of life was needed in all historic times. But at no time were these principles more needed than now. Hence, this is the most appropriate time to take forward the thinking of 'Jai Jagat' with its two core values of 'universe as family' and 'peace and non-violence as a way of life'.

Author's note: This chapter was originally written as a contribution to Jai Jagat Global Peace March. Interested readers can get a more detailed coverage of the issue in my book Earth Beyond Borders.

Holding Hands can Help

A Comprehensive Peace Movement

THERE ARE SEVERAL inspiring and influential peace movements in various parts of the world. These include movements for disarmament, movements for improving relations with hostile countries, movements for ending sectarian violence within a country as well as well as other kinds of peace movements. Some peace movements spring up in response to a new challenge (such as the movement for preventing invasion of Iraq) and then fizzle out while others exist on a more durable basis for bringing peace to some troubled regions. All these are very relevant efforts, but the big question is whether these are adequate.

It is very clear that the destructive capacity of war, weapons and violence is now greater than ever before. While World War II is regarded as the most destructive war in human history, it is only in the period after this that the accumulated weapons acquired the capacity to destroy all human life and most other life forms as well.

It is true that nuclear disarmament efforts led to some reductions of these weapons, but as long as the destructive capacity remains more than what is needed to destroy almost all humanity in a matter of minutes there is hardly any real progress in nuclear disarmament. Also, the threat of actual use which appeared to be reducing in the immediate aftermath of the supposed end of the cold war has again started increasing in recent times. Then there is the threat of tactical nuclear weapons and of depleted uranium weapons and of nuclear weapons falling under the control of terrorists or rogue regimes controlled or heavily influenced by terrorists.

While it is true that a very large number of countries favour nuclear disarmament, the fact remains that as long as none of the nine known nuclear weapon states give up or significantly reduce their nuclear weapons the threat level will not decrease. In addition, it is quite possible that the number of nuclear weapons countries may soon increase even further. Several of the known nuclear countries are hostile to each other. Leaders of some nuclear weapon countries have boasted about using their nuclear weapons to destroy others.

It is well known that the destructive capacity of conventional, non-nuclear weapons is also increasing all the time. There are very powerful interests in the arms industry which push all the time for more and more purchase of more and more destructive weapons. These weapons cause the most terrible destruction and suffering in many parts of the world.

The huge budgets spent on these highly expensive weapons result in denial of basic needs to a significant part of the world population. Modern wars and weapons are also extremely polluting. If weapons and wars can be eliminated both poverty and pollution (including GHG emissions which cause climate change) can be decreased significantly, apart from the main benefits of saving human lives and avoiding a lot of destruction in a more direct way.

Violence in Everyday Life

While these are the big issues for the world peace movement, there is also the need to keep in mind a different kind of violence –the violence in everyday life. In fact, violence in various forms including violent thinking and self-violence is the biggest cause of distress in daily life. While the more extreme manifestation of this is in the form of homicide, suicide, inflicting serious injury, rape etc. there are numerous other incidents and manifestations of violence and violent thinking which may or may not be covered by crime records but which nevertheless cause a lot of pain and suffering and often also leave longer-term scars on life.

As long as these violent tendencies are there, these can easily link up with any prevailing discriminations or biases existing in a society on the basis of gender, religion, caste, ethnicity, or colour to result in various types of gender, communal, ethnic and caste violence. Some of the sectarian violence can become strong enough to lead to civil war-type conditions in some countries. Easy availability of dangerous weapons increases all these risks. When these violent tendencies are strong , these link up with cross-border problems or suspicions to increase the possibility of hostility and eventually war with other countries.

Hence violence at all levels is highly destructive and there are linkages between violence and violent tendencies at various levels.

While the threat of annihilation of most forms of life caused by nuclear and other weapons of mass destruction calls for a most urgent response at this level, most people are too tied up in their day-to-day problems to link up closely on a continuing basis with such an effort. People are likely to be more responsive to checking those forms of everyday violence which more directly impacts and hurts their lives on a daily basis.

Thus, the peace movement which the world needs is a comprehensive movement which covers the big disarmament issues as well as all violence of daily life. Of course, the various relevant peace movements with their more specific objectives should all continue but they should be increasingly informed and conscious about the larger message of the peace movement to reduce violence and violent thinking at all levels.

A well-integrated peace movement will make it possible to get mass support for issues such as nuclear disarmament and this is even possible in nuclear weapon countries.

A peace-based approach to life which is able to increasingly avoid violent urges and thinking will bring a lot of relief in daily life and on this basis this movement can spread fast among ordinary people if guided and planned properly. This can then provide a great opportunity also for making issues like disarmament a mass movement.

Foundation of Peace
Resisting Greed and Dominance

IF A PERSON'S place in history is to be judged by his or her increasing relevance for the future, then Mahatma Gandhi certainly passes this test of greatness with flying colours. His enduring commitment to peaceful methods for fighting injustice in the middle of several difficulties, criticisms and doubts remains one of the most inspiring achievements of a century which witnessed two world wars.

His creativity in visualising satyagraha and related means of non-violent resistance and persisting with these during his lifelong struggles against the biggest imperial force of his times was a very remarkable achievement, particularly when one keeps in mind the lack of previous similar actions or experiences to learn from.

In the middle of much misunderstanding and criticism, Gandhiji remained very firmly a man of peace all his life and his perspective of looking at various issues and problems was essentially the perspective of peace and non-violence. If his commitment to peace and non-violence made him unpopular, then he was quite willing to accept this unpopularity as the price of what he saw as the path of truth. In this, he was very different from most present-day leaders and statesmen.

Another aspect of Gandhi's deep commitment to peace relates to his very sincere and heartfelt respect for various leading religions as well as his deep commitment to inter-faith peace and harmony. This was such a deep commitment for him that he was willing to sacrifice his life for this. When it came to protecting the just interests of a religious minority, he never hesitated, notwithstanding the opposition of the majority community.

If there is an unwavering commitment to peace and to non-violence, then the perception of world's problems changes. This is more needed today than ever before in a world increasingly threatened by nuclear weapons and other highly destructive weapons. The mass distress caused by war and civil strife and terrorism goes on increasing with the increasing availability of more and more destructive weapons. At the same time disarmament and curbs on weapons get less and less attention.

This, moreover, is one among several aspects of the proliferation of violence, including violence in everyday life. Taken together, all incidents of violence in everyday life take on the dimensions of an ever-continuing war. In such a situation, commitment to a life philosophy based on non-violence and peace, tied to resistance to injustice and inequality, can perhaps achieve more than anything else for reducing distress on a continuing basis.

This violence, in turn, is related to tendencies of greed and dominance which have an overwhelming presence in everyday life as well as in conduct of statecraft and conduct of international affairs. Trying to control a bigger share of resources remains an important cause of conflicts, including some of the biggest conflicts. At the same time, greed and the related urge for dominance remain at the root of much of the violence in everyday life. In this context Gandhiji's lifelong commitment to a life of simplicity and his experiments in areas as diverse as food, clothing, education, health, handicrafts, housing and agriculture signify a search for alternatives which can meet the needs of all people on a sustainable basis.

While some leaders try to get short-term popularity by promising a lot to people which cannot be realised or if realised, cannot be sustained in ecological terms, Gandhiji was far ahead of his times in giving adequate attention to sustainability aspects. In various walks of life what he suggested was based on what is more sustainable, what is durable, what can be accessed by all without being destructive to nature or without imposing a heavy burden on nature.

He said that nature has enough to meet the needs of all people but not to meet the unrestricted greed of a much lesser number of people. This is confirmed by detailed studies which reveal that it is the excessively consumerist and extractive life of a pampered minority which has contributed much more to accentuating the environmental crisis. In terms of resolving the ecological crisis, the present-day solutions are more in terms of technological fixes while Gandhiji's thinking based on simplicity, limited needs and economics based on cooperation for welfare of all instead of pursuit of greed and unlimited self-interest can provide more basic solutions.

Gandhiji rejected both capitalism and communism and instead pleaded for an economic system based on a strong sense of ethics which protects livelihoods and meets basic needs of all.

His idea of economic well-being is also very well integrated with protection of environment and sustainability. In a world threatened as never before by ecological ruin and climate change Gandhiji's vision of economic wellbeing that does not endanger environment is all the more relevant now.

There is a need to take forward the essential message of Gandhiji with understanding and dedication among people, particularly youth, while also relating it to contemporary concerns and issues like climate change.

In this effort we must be prepared to answer several questions which critics including youth often raise. In fact, I have also been a critic of Gandhiji on his handling of several issues and events. I have critisised him several times for not initiating a mass peaceful movement for preventing the execution of Bhagat Singh, Sukhdev and Rajguru, in particular and more particularly for exempting all freedom fighters from capital punishment. I have also argued that Gandhiji's treatment of Subhash Bose – particularly after his re-election as President of Congress – was rather harsh.

While taking Gandhiji among youth we should provide room for acceptance of these and other lapses. But to reject the immense contributions and continuing great relevance of Gandhiji merely on the basis of some lapses, including serious ones, would be a great mistake. Gandhiji deserves a very important place in history for his many-sided and enduring contributions, particularly in the most important areas of peace, sustainable development and protection of environment.

Gandhi's Dharma

Sincere Respect for All Religions

AT A TIME when some powerful elements are trying to co-opt some of the ideas of Mahatma Gandhi without giving up their ideology of communalism, we need to recall that Mahatma Gandhi's views on religion had absolutely no place for communalism and instead were based on very deeply felt, very sincere respect for all religions. He went beyond tolerance and pleaded for respect for other religions. It was a matter of faith for Gandhiji that various religions should get equal respect.

> For me the different religions are beautiful flowers from the same garden, or they are branches of the same majestic tree. Therefore, they are equally true, though being received and interpreted through human instruments equally imperfect (*Harijan*, 30.1.37, pp. 407).

'Each religion has its own contribution to make to human evolution', Gandhi wrote. 'I regard the great faiths of the world as so many branches of a tree, each distinct from the other though having the same source' (*Harijan*, 28.1.39 pp. 448).

Gandhiji said, and sincerely believed, that all prayer, in whatever language or from whatever religion it was, was prayer addressed to one and the same God and taught humankind that all belonged to one family and should bear love to one another.

Various religions were like the leaves on a tree. No two leaves were alike, yet there was no antagonism between them or between the branches on which they grew. Even so, there is an underlying unity in the variety which we see in God's creation (*Harijan*, 26.5.46, pp. 154).

There is, therefore, absolutely no reason why anyone while remaining true to one's own religion should not respect other religions. In fact, those who are free-minded and liberal enough to learn from all scriptures (as Gandhi surely was) are the ones who benefit the most.

> I am a believer in the truth of all the great religions of the world. There will be no lasting peace on earth unless we learn not merely to tolerate but even to respect the other faiths as our own. A reverent study of the sayings of different teachers of mankind is a step in the direction of such mutual respect. (In Search of the Supreme, Vol. III, 1962, pp.10.)

> I believe in the truth of all religions of the world. And since my youth upward, it has been a humble but persistent effort on my part to understand the truth of all the religions of the world, and adopt and assimilate in my own thought, word, and deed all that I have found to be best in those religions. The faith that I profess not only permits me to

> do so but renders it obligatory for me to take the best from whatsoever source it may come (*Harijan,* 16.2.34, pp. 7).
>
> India, with its ancient religions, has much to give and the bond of unity between us can best be fostered by a wholehearted sympathy and appreciation of each other's form of religion' (*The Collected Works of Mahatma Gandhi* Vol.V, pp.50).

A Christian friend asked him, 'Would you say then that your religion is a synthesis of all religions?'

'Yes, if you will. But I would call that synthesis Hinduism, and for you the synthesis will be Christianity (*Harijan* ,6.3.37 pp. 27).

In compatibility with his views on equality of all religions, Gandhiji said that the path of progress is to improve the understanding and practice of your own religion and not try to convert others to your own religion.

> 'After long study and experience, I have come to the conclusion that (1) all religions are true; (2) all religions have some error in them; (3) all religions are almost as dear to me as my own Hinduism, inasmuch as all human beings should be as dear to one as one's own close relatives. My own veneration for other faiths is the same as that for my own faith; therefore, no thought of conversion is possible. The aim of the Fellowship should be to help a Hindu to become a better Hindu, a Musalman to become a better Musalman and a Christian a better Christian' (*Selections from Gandhi*, 1957, pp. 258-59).

In Mahatma Gandhi's religion there is enough scope for rational reasoning and social reforms. 'I reject any religious doctrine that does not appeal to reason and is in conflict with morality. I tolerate unreasonable religious sentiment when it is not immoral' (*Young India,* 21.7.20, pp. 4).

Gandhi went one step further on social reforms when he said, 'True morality consists, not in following the beaten track, but in finding out the true path for ourselves and in fearlessly following it' (*Selections from Gandhi*, 1957, pp. 254).

This broad-mindedness extended not just to religious matters but also to matters relating to different cultures, different value-system and worldviews. Gandhi said, 'I do not want my house to be walled in on all sides and my windows to be stuffed. I want the cultures of all the lands to be blown about my house as freely as possible. But I refuse to be blown off my feet by any' (*Young India*, 1.6.21, pp. 170).

Gandhi's emphasis on reason-based approach to religion and his sincere respect for other religions as well as his firm opposition to communalism need to be remembered and better known today when the heritage of mutual respect of various religions is threatened in several contexts.

POLITICAL AND SOCIAL MOVEMENTS

Oh India, dare to be worthy of your Gandhi.

—Pearl S Buck

Mahatma Gandhi has done more than any other person in history to reveal that social problems can be solved without resorting to primitive methods of violence.

—Martin Luther King

In my life, I have always looked to Mahatma Gandhi as an inspiration.

—Barack Obama

His life has inspired me since I was a little boy.

—Dalai Lama

India's Freedom Movement
Struggles and Constructive Activities

A VERY IMPORTANT aspect of the freedom movement of India was that millions of people, men and women, were involved with deep commitment in peaceful, non-violent actions and movements for resisting the highly unjust and exploitative colonial rule. This involvement continued for nearly three decades. Even school-going children were enthused with the spirit of these struggles and created games around these struggles. As they grew up, they participated valiantly in these struggles at a very tender age. This ability to raise the human spirit to such a high level for a long time was a very big achievement.

Another important aspect of the same peaceful freedom movement was that the main political struggle against colonialism was accompanied by a wide range of constructive activities aimed at social reform, reducing poverty and ensuring sustainable livelihoods.

Removing the terrible practice of untouchability and promoting social equality in other ways was an important aspect of this social reform. Establishing and promoting inter-faith harmony in various ways was another important aspect of social reform. Opposing the sale and consumption of liquor and other intoxicants, symbolised by picketing of liquor shops, was important in itself and also because of the support this received from women. Encouraging a more active social role for women was another important aspect and the peaceful freedom movement itself brought forward protesting women on streets in unprecedented ways.

Promotion of khadi (hand-spun, hand woven cloth) and making bonfires of imported cloth was symbolic of wider efforts and thinking to resist highly unfair, imposed colonial trade (and the related destruction of livelihoods) and protect sustainable livelihoods while at the same time resisting indiscriminate mechanisation.

Promoting public hygiene with voluntary work was an important part of the wider efforts to improve health. There was a concern for evolving low-cost health and education systems more in tune with the real needs of India's poorer people.

In almost all these efforts of the non-violent freedom movement, Mahatma Gandhi played a leading role or an important role. This itself is a very great achievement that peaceful resistance to colonial rule, combined with highly relevant constructive activities, maintained continuity for a long time of nearly three decades. There was even greater continuity in constructive activities than in the activities directly related to freedom from colonial rule.

Very valiant and important contributions were also made by other streams of the freedom movement which need to be recognised. For example, the great contribution of Shahid Bhagat Singh and his valiant colleagues can never be forgotten and in fact was recognised by several close colleagues of Gandhiji like Jawaharlal Nehru.

Similarly, the great social reform efforts of Baba Ambedkar and his colleagues should get our respect. Sometimes, these different streams may appear at odds with each other, but in practice, probably the more important role was for one part of awakening and change to add to another part of awakening and change. Different sincere people and movements contribute to resolving and reducing a serious problem in their own way; the contribution of all sincere and well-intentioned efforts should be recognised.

What is truly disturbing is that all the deep concern for peace could not stop the tragedy of the great violence of partition. In the last days of colonial rule, the partition of India claimed around one million lives and permanently uprooted several times more people from their ancestral homes and environs. The Great Bengal Famine during the last years of colonial rule took a toll of perhaps three million lives. These facts continue to disturb us deeply and need to be mentioned here because we need to learn from history.

The peaceful struggles for the freedom of India were great and inspiring in themselves, but these did not have a happy ending, as Gandhiji himself stated several times in his last days. This fact needs to be recorded clearly, for the purpose is not just to celebrate the peaceful struggles but to learn from them for planning all suuch future struggles.

Golden Opportunities Lost

One Peace Struggle Gandhi Missed

THERE ARE CERTAIN moments of history created by unique circumstances which have to be seized upfront. Such moments, if utilised properly, provide huge potential for progress, harmony and justice. But if such moments are lost, the unique opportunities provided by them return rarely, leaving behind a trail of deep regret for missing out on possibilities of great achievements for the betterment of humanity.

Such a moment in the history of India in the previous century exists around the time when three of the bravest sons of India – Bhagat Singh, Sukhdev and Rajguru – were given the death sentence by the highly arbitrary, contrived and manipulated justice (or rather injustice) system of the colonial regime. The extreme injustice involved in this death sentence and the violation of principles of justice in the course of the trial were widely recognised and in fact there was a campaign even in Britain for taking back the death sentence. In India, of course, people from Punjab to Madras were vociferous about this.

Thus, this was just the right time for the Congress-led mainstream freedom movement to launch a huge national-level movement , on the scale of other big movements like non-cooperation and civil disobedience, to demand that this death sentence should be taken back. A prison sentence could have been accepted. A wider demand could have been that death sentence is ruled out for all those who face trial for their involvement in the freedom movement.

Such a demand would have been well within the principles of peace and peaceful struggles rightly emphasised by Mahatma Gandhi. In fact, it has been one of the enduring demands of the peace movement that there should be no capital punishment. If death sentence is to be ruled out for known criminals, then surely there is hundred times more justification for ruling out death sentence for freedom fighters in a colonised nation. Hence, well within the framework of the highly justified principles of peace emphasised by Mahatma Gandhi it was highly justified to launch a huge national movement for taking back the death sentence of Bhagat Singh, Sukhdev and Rajguru.

In early 1931, the nation was ready for this. It is well known that Bhagat Singh Appeal Committes had been set up in all districts of Punjab and people including students were coming out in huge numbers on streets, braving police attacks, to demand the commutation of this death sentence in Punjab and nearby areas. What is not so well known is the fact that the Tamil Nadu Congress Committee had insisted that commuting this death sentence was an essential condition for peace. Thus, public sentiments against this death sentence were surging over vast parts of the country and the conditions were ripe for the Congress to launch or at least announce a huge national movement on this demand of commuting the death sentence.

Of course, the immediate objective of this movement would have been to save three extremely precious human lives. All of these three freedom fighters were extremely brave and capable persons but Bhagat Singh in particular needs to be singled out as someone who in due course would have emerged as one of the stalwarts of the world socialist movement, in addition to his very important national role. Even a cursory look at his achievements at the age of 23 would convince anyone about such future potential of this young man. Millions of people, particularly youths and women had become very emotionally attached to Bhagat Singh and his comrades and felt very strongly about saving their life. So, the immediate aim of saving these precious lives was good enough reason for launching a big movement. In fact, such a movement is likely to have succeeded also in achieving much wider objectives.

Firstly, at this stage of the freedom movement, it was very important to attract more and more people inspired by high ideals and great determination to make the freedom movement a success. Bhagat Singh had understood clearly that a secretive movement has very limited potential for this. This is why, although the colonial regime had failed to catch him, Bhagat Singh decided to come out in the open by indulging in a public act of defiance and then offered himself for arrest. From now on, all his acts would be open to the public to see and this is what he wanted, not for any fame but only to inspire others to join the freedom movement.

The great sacrifices he and his close colleagues like Jatindra Nath Das made in jail, including fasting for over 60 days and enduring torture and beatings with determination, inspired a very large number of people to join and support the freedom movement. Several hundred thousand people had gathered at various places to pay their homage to Jatindra Nath Das when he died after a 63-day fast. Now, if a national movement was launched for taking back this death sentence of these revolutionaries, then many more people would have been inspired to join the freedom movement.

But an even bigger gain would have been to strengthen a wider unity of the Congress and the revolutionaries. Some of the most popular leaders of Congress like Subhash Chandra Bose and Ganesh Shankar Vidyarthi would have been very happy about this, and they would have been able to mobilise a very large number of very enthusiastic and determined people.

This coming together of the two streams would have also avoided those conditions under which Subhash later had to leave India. Above all, the coming together of the two would have also strengthened greatly the forces of communal harmony. The overall impact would have been to achieve independence at an earlier date without riots and partition.

Unfortunately, however, this great opportunity was lost as the Congress failed to launch a big national movement for the commutation of this death sentence. Bhagat Singh, Rajguru and Sukhdev were hanged by the colonial regime on 23 March 1931.

The entire nation rose as one to pay homage to this martyr and to support the cause for which he had sacrificed his life. His funeral procession in Calcutta led by Subhash Chandra Bose was miles long and about 7 lakh people participated. Homage meetings were organised all over the country. So, if at this time, Gandhiji had announced a satyagraha for protecting the rights of political persons in general and more specifically, for demanding that no death sentence be given to a political prisoner, then there would have been a great upsurge of the freedom movement which would have united all streams of the freedom movement.

A huge historical opportunity was thus lost, a great loss from which we should learn.

Views of Tolstoy and Gandhi

Alcohol and the Anti-liquor Movement

> Men who visit our village do not have an idea of the great distress we women suffer day in and out in our village due to alcohol and alcoholism. You must listen to us.
>
> —Woman in Pather village of Saharanpur, in the middle of an anti-liquor movement

BEGINNING WITH THE Gandhian opposition, the anti-liquor movement, coupled with the literacy and adult education movements became some of the earliest post-independence movements.

While the evidence of adverse health impacts of alcohol and tobacco has become increasingly stronger in recent years, despite the strong efforts of the strong and resourceful liquor and tobacco lobbies to spread disinformation, there is less awareness of the wider adverse social impacts of liquor and tobacco as well as the social mobilisation against liquor, particularly by women.

In this context, the ideas of Tolstoy and Mahatma Gandhi deserve wider attention.

In his younger days Tolstoy had consumed a lot of these substances so he was very familiar with them. In his later years, he not only gave up their consumption but became a strong opponent of their use. His views on this subject need to be better known. Going beyond the well-known adverse health impacts of liquor and tobacco, Tolstoy stated that liquor* and tobacco are very frequently used by human beings to remove the inhibitions that most human beings naturally have against committing any serious wrong and unjust act or crime.

To remove this inhibition one of the most common devices human beings use is to smoke tobacco and drink alcohol, he said. Tolstoy gave several examples to make his point but at that time, he did not have a strong data base to support his viewpoints. In recent years a strong data base as well as other evidence has accumulated linking alcohol consumption in particular to a huge number of crimes, sexual crimes and other acts of violence which strongly support the views expressed by Tolstoy.

Mahatma Gandhi strongly opposed the use of liquor, tobacco and other intoxicants. His special contribution here as in many other fields has been to link ideas with actual action. Under his leadership a strong public campaign against liquor became one of the most important components of the constructive programme that became a very important part of the non-violent, peaceful struggle of the freedom movement. One manifestation of this was in the form of picketing of liquor shops with a strong moral appeal to would-be-customers not to purchase any liquor.

---

* The most common definition of alcoholic beverage is a beverage containing at least 0.5 per cent of alcohol by volume. The most common definition of a standard drink is one which contain 10 grams of pure alcohol.

This movement received very enthusiastic support of women. They were always and still are at the receiving end of this particular consumption. The legacy of this can be seen even today in many anti-liquor movements in India which have the most enthusiastic support base of women.

As consumption of liquor has increased rapidly, India has become the most important country responsible for world-level increase in liquor consumption. Governments in this country have been more concerned with collecting the huge revenue generated by liquor consumption. Besides, liquor mafias are known to be close to politicians, and pay huge sums to them and political parties.

Women suffer the most from this and they have been in the forefront of most anti-liquor movements. Mahatma Gandhi continues to be an inspiration source for many of these movements.

Alcohol's Tragic Health and Social Impacts

There is growing worldwide concern at the escalating costs of alcohol consumption not only in terms of health problems and accidents but also violence and social disruption.

1. Highly Adverse Health Impacts

There is more evidence than ever before of very serious health risks from alcohol but despite this, the overall trend in the world from 2005 to 2025 has been of a significant rise in liquor consumption. This is the overall conclusion of the World Status report on Alcohol 2018, prepared by the WHO and released in September 2018.

The per capita per year pure alcohol consumption at the world level increased from 2005 to 2016 and is projected to increase again from 2016 to 2025. However, the world level increase rate is much lower than the very fast rate of growth of liquor consumption in India. The world level increase is driven above all by the rising trend in two highly populated countries, India and China. The rate of increase is higher in India than in China.

In India, the per capita per year pure alcohol consumption increased from 2.4 litres in 2005 to 5.7 litres in 2016, a more than double increase in roughly a decade. This is projected to increase to about 8 litres in 2025 when for the first time since records are available, the average consumption in India at 8 litres is projected to be higher than would be the world average of 7 litres.

According to the WHO Status report, in 2016, the harmful use of alcohol resulted in some 3 million deaths (5.3% of all deaths) worldwide and 132.6 million disability-adjusted life years (DALYs) – i.e. 5.1% of all DALYs in that year. Mortality resulting from alcohol consumption is higher than that caused by diseases such as tuberculosis, HIV/AIDS and diabetes. Among men in 2016, an estimated 2.3 million deaths and 106.5 million DALYs were attributable to the consumption of alcohol. Women experienced 0.7 million deaths and 26.1 million DALYs attributable to alcohol consumption.

This fairly recently-released report tells us that in 2016, of all deaths attributable to alcohol consumption worldwide, 28.7% were due to injuries, 21.3% due to digestive diseases, 19% due to cardiovascular diseases, 12.9% due to infectious diseases and 12.6% due to cancers. About 49% of alcohol attributable DALYs are due to non-communicable and mental health conditions, and about 40% are due to injuries.

As much as 13.5% of all deaths among those who are 20–29 years of age are attributed to alcohol.

Explaining the situation further, this report tells us that the health and social harms from drinking alcohol occur through three main interrelated mechanisms: i) the toxic effects of alcohol on diverse organs and tissues in the consumer's body (resulting, for instance, in liver disease, heart disease or cancer); ii) development of alcohol-dependence whereby the drinker's self-control over his or her drinking is impaired, often involving alcohol-induced mental disorders such as depression or psychoses; and iii) through intoxication – the psychoactive effects of alcohol in the hours after drinking

According to research works consulted for preparing this status report alcohol consumption has been shown to increase the risk of HIV/AIDS by increasing the risk of transmission (resulting from an increased risk of unprotected sex) and by increasing the risk of infection and subsequent mortality from tuberculosis and lower respiratory infections by suppressing a wide range of immune responses via multiple biological pathways, particularly in people who engage in heavy episodic drinking or who chronically consume large amounts of alcohol

The harmful use of alcohol is associated both with an increased risk of acquiring HIV infection and with negative effects on people living with HIV/AIDS in terms of treatment outcomes, morbidity and mortality.

Further, the WHO status report informs us that alcohol consumption has a synergistic effect with viral hepatitis in the progression of liver disease. In addition, alcohol is a well-known causal factor for non-infectious liver diseases, including hepatitis and liver cirrhosis, and the latter is associated with high mortality.

Coming to cancer, this report says that there is an established causal link between alcohol use and cancer development in the oropharynx, larynx, oesophagus, liver, colon, rectum and the female breast . The risks are generally higher for females than males. Even moderate alcohol intake, corresponding to daily consumption of no more than 25 grams of pure alcohol, has been shown to increase the risk of developing female breast cancer.

This report informs us that the causal relationship of alcohol consumption and liver diseases is well established, and alcohol has been shown to have an ability to cause hepatocellular damage through ethanol metabolism-associated mechanisms and malnutrition Alcohol is one of the most frequent causes of liver disease; alcohol-involved subtypes of liver disease include alcoholic hepatitis, steatosis, steatohepatitis, fibrosis and cirrhosis. Acute alcoholic hepatitis and liver cirrhosis are associated with high mortality (which can reach 50% in acute alcohol hepatitis), and the median survival time of patients with advanced liver cirrhosis can be as low as 1–2 years.

Giving an example from China, this report mentions the Beijing 302 Hospital which is a large hospital treating patients from most parts of China, including over 40,000 patients per year with liver disease. Those treated for liver disease at the hospital are thus reflective of trends in liver disease in China. In the period 2002–2013, the distribution of types of liver disease changed at the hospital, with the proportion of alcoholic liver disease (ALD) more than doubling. Throughout this period, most of the patients with ALD (about 98%) were male. A study reporting this remarked that 'the number of patients with ALD is rising at an alarming rate in China'.

The WHO report refers to expansive literature which shows that alcohol intoxication can increase dysphoria, cognitive dysfunction, impulsivity and intensity of suicidal ideation. People have approximately seven times increased risk for a suicide attempt soon after drinking alcohol, and this risk further increases to 37 times after heavy use of alcohol. The alcohol-attributable fraction for suicide was estimated to be as high as 18%. It is also known that the presence of AUDs (Alcohol Use Disorders) at least doubles the risk of having depression.

This report points out that the potential effects of alcohol include impairment in attention, cognition and dexterity (which are important for such activities as driving a car); aggressive impulses and loss of behavioural control (important for criminal violence); and alcohol poisoning (which can be fatal).

Further, this report tells us that alcohol poisoning is a consequence of drinking large amounts of alcohol in a short period of time. Drinking too much in a short period of time can affect breathing, heart rate, body temperature and gag reflex and may result in a coma and death. In comparisons with other psychoactive substances, alcohol is among the most lethal in terms of how close the amounts used for psychoactive effects are to the median amount that is lethal. While poisoning clusters from contaminated alcohol often receive global press coverage, poisoning with ordinary beverage alcohol – usually in concentrated form such as distilled beverages – is an everyday reality in many societies, although it is often under-recorded in health statistics.

The harmful use of alcohol is a component cause of more than 200 diseases and injury conditions, the most notable being alcohol dependence, liver cirrhosis, cancers and injuries, the latest causal relationships established are those between alcohol consumption and incidence of infectious diseases such as tuberculosis and HIV/AIDs.

However, the alcohol industry and its lobbyists have worked overtime to ensure that the massive adverse impacts of alcohol consumption are not reported adequately and properly so that people do not become aware about the full dimensions of these tragic impacts.

A recent study led by scientists at the London School of Hygiene and Tropical Medicine and Sweden's Karolinska Institute has concluded that the alcohol industry 'uses denial, distortion and distraction to mislead people about the risks of developing cancer from drinking, often employing similar tactics to those of the tobacco industry'. However, the WHO says that drinking alcohol is a well-established risk factor for a range of cancers including tumours of the mouth, liver, breast, colon and bowel, and the risk of cancer rises with the level of alcohol consumed.

Earlier also, heavy drinking was linked to damage to brain, adverse impact of memory and dementia, but recent findings by researchers of Oxford University and University College London (published in British Medical Journal) has found that this damage is possible also at much lower alcohol consumption. This is also confirmed by another study involving 1,300 women in the USA. Brain damage is likely to be higher in the case of binge drinking, particularly binge drinking involving adolescents.

Although the highly adverse impacts of alcohol consumption are very well-known, the alcohol industry has been trying for quite some time to somehow spread the myth that moderate levels are not a problem.

However, an extensive study based on 195 countries which was published in the prestigious medical journal The Lancet (August 2018) concluded clearly that not even one drink a day is safe.

Max Griswold, lead author for a group of over 500 experts said, 'There is no safe level of alcohol', 'Overall, the health risks associated with alcohol rose in line with the amount consumed each day'.

Compared to abstinence, just one drink a day can lead to 100,000 additional deaths each year.

In the 15-49 age-group, alcohol is responsible for over 12% of deaths among all.

Deaths in this age-group is generally believed to cause more distress in households.

Considering 24 health problems, this study found that five drinks a day can raisesevere health consequences by 37%. In Romania male drinkers average 8 drinks a day, while in Portugal and Ukraine they average 7 drinks a day.

2. Accidents

According to the Encyclopedia of Drugs, Alcohol and Addictive Behaviour, alcohol has been found to have a role in 44 per cent of fatal road accidents. Up to 50 % per cent of motorcyclists who crashed to death are likely to have been under the influence of alcohol.

Regarding non-fatal accidents, this encyclopaedia tells us that alcohol is involved in 23 to 30 per cent of these accidents. In the case of fatal fire and burn accidents, alcohol was found to have a role in 46 per cent of such accidents.

3. More Serious Health Impacts

Less than two years after these were issued by a determined woman medical official with the support of senior cancer experts, new British guidelines on liquor are being seen as a big help by campaigners against the many-sided risks of alcoholic drinks not just in Britain but in other countries as well.

Dame Sally Davies, Chief Medical Officer of Britain, overcame a lot of resistance from liquor industry lobbyists to radically change two-decadeold guidelines which had understated health risks of liquor while also either implicitly conceding or at least not adequately refuting imaginary claims of some health benefits (for example in the context of red wine) spread by lobbyists and sellers of liquor.

The new guidelines came in the wake of growing concern over reported 500 per cent increase in deaths from liver disease among working age people in Britain since the 1970s. Another serious concern came with growing evidence of higher risks of a range of cancers associated with liquor, including red wine.

The evidence regarding the closer link between liquor and a range of cancers was supported by a new review from the Committee on Carcinogenicity in Britain.

The development of the new guidelines was chaired by Prof Mark Petticrew, Professor of Public Health Evaluation at the London School of Hygiene and Tropical Medicine and by Prof Sally Macintyre, Professor Emeritus at the University of Glasgow.

The new guidelines on liquor consumption released on 8 January 2016 by the Chief Medical Officer to replace nearly two-decade old guidelines warn that drinking any level of alcohol increases the risk of a range of cancers. The guidelines also make it clear that no level of alcohol is safe for drinking in pregnancy.

These guidelines inform that drinking regularly over time can lead to a wide range of illnesses including cancers, strokes, heart diseases, liver diseases, and damage to the brain and nervous system.

On the other hand, these guidelines also make it clear that there is no justification for drinking for health reasons.

Infact, the highly exaggerated and mostly imaginary health benefits cited to justify increased consumption of red wine had led to increased chances of addiction among a different category of drinkers who had been careful regarding hard liquors like whisky, vodka and rum but had become careless regarding consumption of red wine and fortified wines due to the false myths spread regarding their lower risks and higher benefits of red wine.

The alcohol content of red wines and fortified wines is between 10 to 22 per cent for various brands. This is three or more times higher than the alcohol content of beer which ranges between 2 to 10 per cent. Despite this red wine was promoted as a low risk liquor and at times this was promoted as a health drink which benefits heart.

Later, even some of those who claimed such benefits had to admit that these could be more easily obtained from some raw grapes and berries and from some types of exercise and there was no need at all to accept the various risks of alcohol consumption to get these benefits.

However, as the myth about benefits and low risk had been spread widely, a significantly large number of people started drinking red wine as a matter of almost daily habit after returning from work. It was against this background that a huge increase in liver disease deaths among working age people caused a lot of concern. At this stage, health professionals and activists started warning about this. Duncan Selbie, head of an organisation Public Health, England presented in his calculation of health impacts that one glass of red wine is the equivalent of three shots of vodka. This was widely reported.

Around the same time, some objections were also voiced by environmentalists regarding excessive use of agri-chemicals and water by the wine industry and also regarding its high GHG emissions. The Economist reported that 960 litres of water is required to make available one litre of wine. To produce 274 million hectolitres of wine in a year, the amount of water needed can be well imagined. About 19 million acres have to be devoted to grow the essential raw material for the wine industry. Hence a lot of precious land and water are taken away from producing staple food at a time when food security is increasingly threatened.

All these factors taken together can lead to a better comprehension of the real costs of the liquor industry.

As we in India are painfully aware, the costs of high liquor consumption in terms of road accidents and other accidents, ruin of health, economic loss, domestic violence, sexual violence and other violent crime are very high. Despite the availability of shocking facts and figures on all these aspects, this issue does not get the attention it deserves among elites.

However, people, particularly women at many places are fighting to reduce the consumption and sale of liquor in various ways. Efforts such as the issue of new guidelines in Britain are a big help for creating conditions in which the voices of these struggles can be heard.

4. Depression and Suicide

The World Report on Violence and Health (WRVH) says that alcohol abuse may also be an important factor in depression. More ominously the WRVH report says that alcohol and drug abuse also play an important role in suicide. In the USA, at least one quarter of all suicides are reported to involve alcohol abuse.

According to the American Psychological Association, 'Alcohol use and alcoholism can worsen existing conditions such as depression and induce new problems such as serious memory loss, depression or anxiety'.

According to NIH, USA, 'Alcohol abuse may lead to suicidality through disinhibition, impulsiveness and impaired judgment but it may also be used as a means to ease the distress associated with an act of suicide'.

Researchers who compared alcoholics with non-drinkers found suicide tendencies in the former to be 60 times or more. Research by Razvodovsky in Russia revealed that suicides decreased in areas where vodka consumption was restricted.

5. Domestic Violence

According to the WRVH which examined a lot of the available studies on domestic violence, 'the evidence is that women who live with heavy drinkers run a far greater risk of physical partner violence, and that men who have been drinking inflict more serious violence at the time of an assault'. More specifically, this report mentions a survey in Canada which revealed that women who lived with heavy drinkers were five times more likely to be assaulted by their partners than those who lived with non-drinkers.

6. Sexual Violence

The WRVH says that both from the perspective of the assaulter and the victim, alcohol and drug consumption increases the risk of sexual violence, including rape. In the context of the victim this report says that consuming alcohol or drugs makes it more difficult for women to protect themselves 'by interpreting and effectively acting on warning signs'. In the context of the assaulter, this report says that alcohol has been shown to play a disinhibiting role in certain types of sexual assault.

According to a widely cited paper on 'alcohol and sexual assaults' by Antonia Abbey, Tina Zawacki and others of the National Institute of Alcohol Abuse and Alcoholism (USA), 'atleast one half of all violent crimes involved alcohol consumption by the perpetrator, the victim or both. Sexual violence fits this pattern. Thus, across disparate population studies, researchers consistently have found that approximately one half of all sexual assaults are committed by men who have been drinking alcohol. Depending on the sample studied and the measure used, the estimates for alcohol use among perpetrators have ranged from 34 to 74 percent. Similarly, approximately one third of all sexual assault victims report that they were drinking alcohol at the time of assault with estimates ranging from 30 to 39 per cent'. However, these researchers also point out that while a woman's alcohol consumption may place her at increased risk of sexual assault, she is in no way responsible for the assault. The researchers rightly say that the perpetrators remain legally and morally responsible for their behaviour.

However, it also needs to be pointed out that while the data about alcohol consumption by perpetrators is relevant to almost all situations, the data about alcohol consumption by victims may not be relevant to those social contexts such as several parts of India where alcohol consumption by women is very low.

In a widely quoted paper, Martin and Hummer argued that many fraternities create a social environment in which social coercion is normalised because women are perceived as commodities available to meet men's sexual needs. Alcohol is used to encourage reluctant women to have sex. One fraternity man quoted in this paper said, 'We provide them with "hunch punch" and things go wild. We get them drunk and most of the guys end up with one'. He describes his plan to get one 'prim and proper sorority girl' get drunk by spiking her punch.

According to another paper by Antonia Abbey, 'the peer norms for most fraternity parties are to drink heavily to act in an uninhibited manner and to engage in casual sex'.

According to a recent report from the National Task Force on College Drinking (USA), 1,400 college students die each year in alcohol-related accidents, 5, 000,00 are injured and there are 70, 000 victims of alcohol-related sexual assault or date rape.

A study by Testa and Livingston mentioned women's narrative description of incapacitated rape which indicates that many were unconscious and found out later that they were raped, or else were only dimly aware of what was happening and so were unable to stop the assailant.

7. Special Vulnerability of Women

New research has not only confirmed special vulnerability of women to alcohol but in addition, the high risks to babies in case of consumption of alcohol during pregnancy are more widely recognised now. The latest Status Report on Alcohol and Health released by the WHO very recently in September 2018 has thrown important light on several important aspects of these issues.

It is widely recognised that alcohol has extremely serious adverse health impacts. However, what is not so widely realised is that women are more vulnerable to many-sided health hazards of alcohol. According to the WHO, 'there is evidence that women may be more vulnerable to alcohol-related harm from a given level of alcohol use or a particular drinking pattern. The vulnerability of females to alcohol-related harm is a major public health concern because alcohol use among women has been increasing steadily in line with economic development and changing gender roles and because it can have severe health and social consequences for newborns.... For health outcomes such as cancers, gastrointestinal diseases or cardiovascular diseases, the same level of consumption leads to more pronounced outcomes for women.'

The vulnerability of women may be explained by a wide range of factors. For example, women typically have lower body weight, smaller liver capacity to metabolise alcohol, and a higher proportion of body fat, which together contribute to women achieving higher blood alcohol concentrations than men for the same amount of alcohol intake.

Women are also affected by interpersonal violence and risky sexual behaviour as a result of the drinking problems and drinking behaviour of male partners. Moreover, alcohol use has been shown to be a risk factor for breast cancer. Also, many societies hold more negative attitudes towards women's drinking alcohol than men's drinking, which, depending on the cultural context, may increase women's vulnerability to social harm. Finally, women who drink during pregnancy may increase the risk of fetal alcohol spectrum disorder (FASD), and other preventable health conditions in their newborns.

The WHO Status Report on Alcohol and Health 2018 has clearly stated that alcohol use during pregnancy has been established as a risk factor for adverse pregnancy outcomes, including stillbirth, spontaneous abortion, premature birth, intrauterine growth retardation and low birth weight , and can result in a range of lifelong conditions known as fetal alcohol spectrum disorders (FASD).

As pointed out in this WHO Report, FASD is an umbrella term which includes fetal alcohol syndrome (FAS), partial fetal alcohol syndrome (PFAS), alcohol-related neurodevelopmental disorder (ARND) and, depending on the diagnostic and classification system, alcohol-related birth defects (ARBD) FASD is characterized by central nervous system damage and its manifestations include physical, mental and behavioural features and learning disabilities with possible lifelong implications.

Presenting a review of the available research on this issue, the WHO staus report states: A systematic review and meta-analysis by Popova and colleagues estimated that the global prevalence of alcohol use during pregnancy in the general population amounts to 9.8% . In addition, Lange and colleagues observed, at the country level, that binge drinking during pregnancy ranged from 0.2% to 13.9% (Note: health risks from binge drinking can be much higher than from normal drinking). Drinking during pregnancy resulted in an estimated prevalence of FAS in the general population of 14.6 per 10 000 people and a prevalence of FASD of 77.3 per 10, 000 people .

The prevalence of FAS and FASD varies by WHO region, with the prevalence being highest in the Region of the Americas (AMR) and the European Region (EUR) . Additionally, a recent systematic review and meta-analysis by Popova and colleagues found that 428 conditions co-occurred with FASD . Some of the most common health problems are congenital malformations, chromosomal abnormalities, prenatal and postnatal growth delays, intellectual disability, behavioural disorders, speech and language difficulties, visual and audiological impairments, cardiac deformities and urogenital problems.

In some cultures, women have been protected from alcohol consumption by strong traditional values.Imagine the public health and social disaster that could result if alcohol consumption by women in India reaches the level of the USA. Yet, promoting alcohol consumption among women as a sign and symbol of liberating modern trends is widely practised in India and the alcohol industry puts in a lot of effort and money to promote this view. Some recent Bollywood movies which promoted this view in a big way were widely praised by big media as a liberating trend. Such perceptions should be corrected before it is too late.

8. Dangerous trend of 'Drunkorexia'

In recent years, a new dangerous trend has emerged in several Western countries, particularly in the college campuses of these countries. Young women (and to a much lesser extent young men also) have been skipping meals to reduce their calorie intake so as to make room for calories they derive from alcohol consumption. This is happening despite the well-known fact that consuming alcohol on an empty stomach is more dangerous. This trend is called 'Drunkorexia'.

Recently the prestigious UK newspaper, The Independent quoted a regular practitioner of drunkorexia saying, 'all alcohol contains loads of empty calories, yet we are also expected to be slim. I and others I know do as many shots of vodka as we can so we get drunk on the least amount of calories and make bloating less likely'.

The Independent also quoted a study which revealed that almost sixty per cent of female under graduates exhibit drunkorexic tendencies.

According to the Eating Disorder Center of Denver, of the participating college-aged females in an adjunct research study, about 75 per cent met the criteria for alcohol abuse.

In a study in Australia involving women aged between 18 to 29 years, who were enrolled in university, 79 per cent of participants demonstrated engaging in characterized drunkorexia behaviour.

Drinking on an empty stomach makes it possible for ethanol to reach the blood system at a faster pace and to raise a person's blood alcohol content at a high- risk speed. This can increase the chances of alcohol related brain damage. In addition, drunkorexia also increases the possibility of black outs, alcohol poisoning, injury and illness. These harmful impacts are more likely in the case of women compared to men as women are generally less capable of metabolizing alcohol than men.

In some cases drunkorexia also involves self-induced vomiting soon after the consumption of alcohol.

9. Youth

Recent research has drawn attention to evidence from several parts of world regarding increasing and many-sided health, learning capacity and safety risks to students and youth from alcohol .The latest available data for 2016 given in the WHO Status Report on Alcohol and Health released in September 2018 tells us that worldwide, more than a quarter (26.5%) of all 15–19-year-olds are current drinkers, amounting to 155 million adolescents.

Further this report says that results of school surveys indicate that in many countries of the Americas, Europe and Western Pacific alcohol use starts before the age of 15 years and prevalence of alcohol use among 15-year-old students can be in the range of 50–70% with remarkably small differences between boys and girls.

Summarising the result of several studies, this report tells us that alcohol use in adolescents is associated with alterations in verbal learning, visual-spatial processing, memory and attention as well as with deficits in development and integrity of grey and white matter of the central nervous system. These neuro-cognitive alterations by adolescents' alcohol use seem to be related to behavioural, emotional, social and academic problems in later life, the report adds.

Some problems are also passed on to children and youths from alcoholic parents, particularly father. Parental drinking or poor parent-adolescent relationship quality are among the factors related to adolescent alcohol use, this report says quoting several researchers.

As this report points out, there is a consistent and large amount of evidence demonstrating the positive association between parental and offspring drinking. Parental alcohol problems were associated significantly with a higher likelihood of frequent emotional symptoms, depression, low self-esteem and loneliness among both boys and girls. In addition, parental heavy drinking and alcohol problems were shown to be associated with worsening outcomes in children, including educational outcomes, drug use, conduct problems and criminality. Parental alcohol use disorders (AUDs) are associated with a higher risk of their children developing depression in adulthood.

According to the WHO, alcohol consumption in adolescents, especially binge drinking, negatively affects school performance, increases participation in crime and leads to risky sexual behaviour. The WHO quotes growing scientific evidence that has demonstrated the special harmful effects of alcohol consumption prior to adulthood on brain; mental, cognitive and social functioning of youth and increased likelihood of adult alcohol dependence and alcohol related problems among those who drink before physiological maturity.

The World Report on Violence and Health(WRVH) says especially in the context of youth violence that drunkenness is an important immediate situational factor that can precipitate violence. In a Swedish study on youth-violence, about three-quarters of violent offenders and around half the victims of violence were intoxicated at the time of the incident.

Several surveys indicate high and rapidly increasing consumption of alcohol amongst youth. What is particularly worrying is high consumption at a very tender age which is likely to be very harmful for brain and also lead to much higher risk of alcohol dependence in later years. For example, in UK binge drinking thrice or more a month is reported among 33 per cent of adolescents in the 15-16 age groups. Binge drinking levels affecting between 25 to 40 per cent of youth are quite common now.

Coming now to the situation in India the WHO status Report on Alcohol and Health has said that India has emerged as a country which is recording some of the highest increase in liquor consumption in the entire world. Although it is not specifically mentioned in this report to what extent this increase in liquor consumption has also led to an increase among students and youth there is a lot of sporadic evidence from many parts of India which indicates that the increase of alcohol among youths in India may be even higher than the overall trend of increase in liquor consumption in the country.

In India

In many parts of the country which this writer visited to write about anti-liquor movements, it was generally seen that women were taking the lead in these movements and their biggest grievance was that their adolescent children are also falling prey to the liquor habit due to the overall increasing trend of liquor consumption and the location of liquor vends within the village.

Heavy episodic drinking, often called binge drinking, is generally regarded as a riskier form of liquor consumption. The WHO Status Report defines this as the consumption of at least 60 grams or more of pure alcohol on at least one occasion in the past 30 days. In India in the 15-19 age group 12.6 per cent of adolescents had indulged in this (21.1 per cent of males and 3.6 per cent of females).

Clearly, there are several very worrying aspects of these trends and we need to wake up to these problems and carry out special campaigns among youth and students before it is too late.

10. Elderly

Among elderly people the possibility of alcohol reaching and affecting sensitive organs including brain, liver and muscles is higher compared to younger people.

11. Crime and Violence

The WRVH report says that reducing the availability of liquor can be an important community strategy to reduce crime and violence as research has shown alcohol to be an important situational factor that can precipitate violence.

In a 4-year study conducted in New Zealand, crime rates in situations of high and low availability of alcohol were compared. This study revealed that crime rates fell significantly for 2 years in areas of reduced alcohol availability.

According to a report prepared by the National Council on Alcoholism and Drug Dependence, USA (NCADD), alcohol and drugs are implicated in estimated 80 per cent of offences leading to incarceration in the USA such as domestic violence, driving while intoxicated, property offences, drug offences and public ordered offences. Alcohol is a factor in 40 per cent of all violent crimes and according to the department of justice, 37 per cent of about 2 million convicted offenders currently in jail report that they were drinking at the time of the event. Alcohol, more than any illegal drug, was found to be closely associated with violent crimes including murder, rape, assault, child and spousal abuse. About 3 million violent crimes occurred each year in the USA in which victims perceived the offenders to have been drinking and statistics related to alcohol use by violent offenders show that about half of all homicides and assaults are committed when the offender, victim or both have drinking.

This report by NCADD points out that alcohol is often a factor in violence where the attacker and the victim know each other. Two thirds of victim who were attacked by an intimate (including a current or formal spouse, boyfriend or girlfriend) reported that alcohol had been involved.

More than 1 million are arrested for driving while intoxicated in a year in the USA. Drinking and drugged driving is the number one cause of death, injury and disability of young people under the age of 21.

12. Impact of Alcohol Consumption on Others

Most research on impact of alcohol consumption is concerned with the impact on the actual consumer, but more recently there has been a welcome trend to look at the impacts on others. The most discussed study of this kind was conducted in Australia and has been mentioned as a very significant study by the WHO which is trying to take this further. This study supported by the Foundation for Alcohol Research and Education found the following impacts in one year in Australia.

(i) 367 persons were killed
(ii) 13,660 persons were hospitalised
(iii) Special care had to be arranged for 19,443 children
(iv) 25,481 family members were attacked
(v) 44,852 strangers were attacked
(vi) 73 per cent persons were affected in some way or the other to a lesser or greater extent

Beyond causing harm to the drinker, alcohol intoxication is often the main mechanism of harm to others from drinking.

Studies of drinking by the offender and by the victim in homicides have chiefly been in societies where drinking to intoxication is common. Meta-analyses of these studies have found that 48% both of victims and of perpetrators had been drinking when the homicide occurred, and that 37% of the offenders and 33–35% of the victims had drunk enough to be intoxicate.

13. Overall Social Impact

There are very heavy social costs of alcohol consumption in the form of various forms of violence, crime and disruption of relationships. The WHO status report has presented various estimates that have been made from time to time about financial implications of these social costs:

(i) Estimate for the European Union, Year 2003-125 billion Euros
(ii) Estimate for UK, Year 2009-21 billion pounds
(iii) Estimate for USA, Year 2006-233 billion dollars
(iv) Estimate for South Africa, Year 2009-300 billion Rand (10-12 per cent of GNP)

While this data about financial implications of social costs is very disturbing, at the same time it needs to be emphasized that violence leading to loss of life and disruption of close relationships caused by alcohol can never be fully calculated in money terms; it goes beyond that.

14. Increasing Consumption in Developing Countries

The WHO status report points out that the most prevalent tendency worldwide is increase in recorded per capita alcohol consumption. Further this report says that while per capita consumption of alcohol at present is highest in developed countries like the USA and European countries, the trend of increasing consumption is mainly driven by increase in alcohol consumption in China and India and this in turn is driven by the active marketing by the alcohol industry.

Another cause of concern is the increasing consumption of liquor by women in several countries despite the fact that alcohol has some special health risks for women. For example, fetal alcohol syndrome and fetal alcohol effects, preventable causes of mental retardation may result from alcohol consumption during pregnancy.

15. Environmental Ruin

Very high levels of ecological ruin and greenhouse gas emissions are involved in the liquor industry. Several water sources including rivers are heavily polluted. A lot of water is wasted. The Economist reported some time back that 960 litres of water are used to produce one litre of wine.

16. Food Security

In order to get the raw material for the liquor industry a lot of land is diverted from food crops, or else some crops are diverted from food to alcohol due to high profit motive. A lot of irrigation water is also diverted from supporting nutrition to supporting intoxication.

17. Policy Initiatives to Check Alcohol Consumption

The many-sided devastation of health and social relations as well as the huge number of injuries and violent episodes caused by alcohol consumption and addiction, as also the huge economic and ecological costs associated with this have been well documented at the world level. In developing countries like India there are three factors that accentuate this loss.

Firstly, many of the families affected are already so poor or economically stressed that the recurring expenditure on liquor simply ruins them. Secondly the poor availability of medical and de-addiction facilities means that the way out of the morass is available to only a few who need this desperately. Thirdly the gender equation is such that despite very low levels of liquor consumption by women it is women and children who end up bearing the heaviest burden of increasing liquor consumption in terms of denial of basic needs, increasing domestic and sexual violence.

Political Nexus

It is in this context that the demands for curbs on increasing liquor sale and consumption have been raised by several anti-liquor movements in various parts of India, particularly rural India. These include social movements with a larger agenda which have also felt the need for resisting liquor related problems and also very specific anti-liquor movements focusing mainly or only on this issue. Women have played the most important role in most of these movements and have brought a strong sense of urgency and passion into these movements. One important specific demand of several of these movements has been to remove the liquor vend or shop from their village so that easy availability of liquor within the village or very near to the village can be checked.

While most governments have paid lip service to anti-liquor sentiments, in reality most governments have been happy to collect their massive earnings from sale of liquor. In addition, several politicians have close links with liquor mafias and some of them have themselves entered liquor production and trade. Despite all this, several governments have been forced to respond to growing public sentiments against liquor, particularly in the context of women voters. They have responded to the anti-liquor sentiments of people particularly women in various ways. Prohibition has been in existence in Gujarat and parts of North-East India and was more recently introduced in Bihar. Tamil Nadu and Kerala are likely to move towards phased prohibition. Other State governments have agreed at times to removing liquor vends if certain conditions such as opposition by more than 50 per cent local population are satisfied and proved. Others have withdrawn specific vends following opposition.

On the whole, however, the massive, many-sided damage caused by alcohol has been increasing. However,r one ray of hope is provided by the fact that where sustained efforts by people's movements are made there alcohol consumption comes down significantly and this reduction is sustained for a long time.

One example is the anti-liquor effort associated with the Chattisgarh movement of iron ore miners led by Shankar Guha Niyogi where several thousand miners gave up liquor motivated by women as well as by wider trade union loyalty and commitment. The big lesson here is that sustained public campaigns and community efforts are necessary for sustainable reduction of alcohol consumption and other substance abuse.

Merely government action cannot succeed in a significant and sustainable way even if it is well-intentioned, this has to be supplemented by a sustained, continuing public campaign against liquor and other intoxicants and various forms of substance abuse. This should involve various respected members of community who have credibility. This campaign should have strong links with schools and colleges so that a strong anti-liquor message can be imparted in friendly and interesting ways at an early age and can be included in the curriculum.

Thus, a combination of suitable, well-thought government policies suited to local conditions and sustained, broad-based public campaigns with women and respected community leaders playing an important role can together succeed in reducing liquor consumption and all the associated problems to a considerable extent.

18. Strong Anti-liquor Policy Needed

There should be a clear and consistent policy guideline that alcohol consumption needs to be minimized and discouraged as much as possible. There is adequate evidence of many- sided serious adverse impacts of alcohol consumption to support such a policy. Although efforts are being made all the time by the alcohol industry and its lobbyists, as well as many layers of liquor mafia in many countries including India, to dilute the evidence against alcohol and even promote its imaginary benefits, such misrepresentations of facts and analysis should not be allowed to dilute any strong policy decisions to discourage and minimise consumption of liquor.

What exact form this policy takes will depend on diverse conditions of various societies. What is perhaps common to all societies is that there should be a well-informed sustained campaign against the many-sided adverse impacts of alcohol. Such campaigns should be adequately supported by governments. All containers of alcoholic drinks should carry strong warnings of health and social adverse impacts. Efforts to present alcohol in attractive settings should be discouraged in various ways. There should be a ban on advertisements of alcoholic drinks.

More specifically, in conditions of rural India – there should be clear policy that no liquor vends or shop can be opened in a village where 50 per cent or more people sign a petition against it. At the same time, there should be very strong action against illegal or secretive sale points of liquor.

Similar decisions can be taken for urban localities keeping in view their specific conditions.

The government policy should cease to see liquor as a source of obtaining more and more revenue. Instead, looking at the wider, much higher adverse impacts of liquor the government should make available resources for strong action to minimize the consumption of liquor as well as for efforts of de-addiction and rehabilitation.

The government should try to take the active participation

of people, particularly women, in many-sided efforts to curb liquor consumption. Once such widespread public participation is ensured, using legal efforts to curb liquor consumption in various ways will also be more successful.

19. Diverting Tea and Fruit Production in Himachal Pradesh

Very worrying trends have been reported recently from Himachal Pradesh – that highly nutritious fruits are being diverted from direct consumption to the production of wine on a large scale. Such trends, if left unchecked, can easily spread to other parts of the country.

First a leading newspaper, *Dainik Jagran* reported that apples as well as other highly nutritious fruits in the State will be diverted on a large scale for production of wine and the existing processing facilities of a public sector unit will be made available to a private company for taking this new wine to many parts of country. Then, on 1 May, 2018 another news appeared in another leading newspaper, *Navbharat Times* that along with fruits, special Kangra Tea leaves will be used to prepare 'tea wine' with alcohol content of 12 per cent (much higher than beer). A government science lab is being used to prepare this and a private company has been contracted to sell this as tea-wine on a large scale. The government will also be earning from this deal.

It is very worrying that even such a staple drink as tea which is consumed by an overwhelming majority of people in the country is being used to prepare an alcoholic drink based on tea leaves with a high alcohol content and at the same time fruits are being diverted for the production of this wine, at a time when children from poor families in this apple-producing State find it very difficult to eat apples.

It is likely that soon there will be a drive by unlicensed shops to sell these new forms of liquor, as can already by seen as some general merchants and fruit juice shops are seen selling fruit wine. What is even worse is that high alcohol content drink or liquor is being promoted for health benefits, a trend which has already been exposed for all its falsehood, deceit and pressures of liquor lobby in the context of red wine.

It is now well known that the completely false promotion of red wine as a healthy drink led to massive damage to health and a huge increase in liver disease as well as other serious health problems. The same tragedy will be repeated in India on an even larger scale when fruit wine and tea wine are promoted for their false health benefits while covering up the conspiracy of the liquor lobby in taking liquor to more and more people.

Hence, this dangerous trend and the distorted thinking behind it should be checked as early as possible.

20. Powerful Interests

While a number of recent research reports have drawn attention to health risks from alcohol which are much higher than what was widely believed a few years back and also to very devastating social impacts of alcohol, this has not led to the necessary public action in India to significantly reduce liquor consumption due to lobbying by highly powerful alcohol industry interests.

It is widely realised that more efforts to curb liquor consumption are needed at the level of various nations which should also have much stronger international support. The recently released World Status Report on Alcohol and Health 2018 prepared by the WHO has also argued along similar lines. This report says, 'At the international level, the broad scope and magnitude of health and social problems caused by the harmful use of alcohol require coordinated and concerted actions by different parts of the United Nations system and regional intergovernmental organizations in the context of the 2030 Agenda for Sustainable Development'.

The Status Report has pointed out that neither alcohol nor tobacco was included in the modern international drug control treaties when they were adopted during 1961-1988. With development and ratification of the Framework Convention on Tobacco Control (FCTC) in response to the globalisation of the tobacco epidemic, now 'alcohol remains the only psychoactive and dependence-producing substance with significant global impact on population health that is not controlled at the international level by legally-binding regulatory frameworks'. Clearly, strong international action to check alcohol consumption is overdue.

The latest WHO status further draws attention to the fact that as a consumer product and trade item, alcohol comes under the jurisdiction of international trade treaties and the international trade system, which have focused on maximising the freedom of international trade and investment.

These international trade treaties, therefore, minimise 'technical barriers to trade and other alcohol control measures' that national or local governments may impose. While there are 'public health' exceptions in the treaties, the WHO report points out, these are interpreted narrowly, so that a provision in the interest of public health is not accepted if an alternative measure can be put forward as an option, whether or not the measure would be effective.

Hence, the WHO Status Report on Alcohol and Health states very clearly:

> Given the increasing encroachment of international trade and investment on the ability of national and sub-national governments to control their alcohol markets, and the increasingly consolidated global alcohol producers, a strong public health case for considering the negotiation of a Framework Convention on Alcohol Control, or alternatively for including alcohol within other international control systems has been repeatedly presented and discussed by public health entities, experts and advocates.

This report quotes studies to make the point that alcoholic beverage production has become increasingly concentrated and globalised in recent decades, particularly in beer and spirits, but increasingly also in wine. An additional adverse factor from the point of view of health impacts is that very significant proportions of the commercial alcohol market are consumed in heavy drinking occasions, meaning the 'central interests of alcoholic beverage producers are inherently at odds with the public health interest'.

This report points out that significant influence of alcohol industries on political decisions which affect them has been documented at national and sub-national levels. In some countries, with substantial alcohol production and trade, this influence extends to the international level, both through national delegations in international organisations and negotiations, and through national, regional and global alliances within and across industries.

What is more, this report states, this interference was reported at several levels, including industry presence during legislative meetings as well as challenges to the legal basis of existing alcohol policies.

Despite this troubling evidence, the WHO report regrets that there is no public health-oriented international agreement on alcohol to counter the strong influence of the producers through the international trade institutions and agreements.

On the contrary, the status report tells us, general trends towards deregulation in recent decades have often resulted in the weakening of alcohol controls – to the benefit of economic interests but 'at the expense of public health and welfare'.

Keeping in view all these adverse factors, this report asserts that substantial added resources are needed to provide leadership at the international level to pursue the public health agendas for alcohol in the face of the strong presence of the global industry at the international policy level.

This report refers to studies which tell us that alcohol producers may also influence public health and welfare perspectives in research. A glaring example given in the status report is that a study of potential health benefits of light drinking (one drink per day) versus abstention, with alcohol industry multinationals and alliances as the main sources of the US$ 100 million funding, was discontinued after a government review group concluded that the interactions of the study's investigators 'and the alcohol industry "appear to have intentionally biased the framing of the scientific premise of the study" in the direction of focusing on possible benefits of alcohol'. This is by no means an isolated example.

The status report informs us that such efforts to bias research and public health policies to accommodate industry's interests, often with as little record of the influence as possible, can also be found in many countries and situations

In India, very powerful and resourceful lobbyists have been at work overtime to increase alcohol consumption even in the remotest villages while women in distant villages have been struggling to stop the rapid increase of liquor consumption. We need to stand on the side of these brave women and support them. This Gandhian struggle needs to go on.

Land Struggles and Justice
Initiatives of Ekta Parishad

AT A TIME when land-reforms, and in particular the land rights of the poorest landless people, are being increasingly pushed to margins, it requires a lot of courage and a firm faith in one's convictions for any organisation to pour all its efforts and energy into a nationwide movement for land reforms. But this is precisely what Ekta Parishad has done, organising a continuing series of foot-marches in 338 districts, an effort that continued for one year (Oct 2011 to Oct 2012).

Ekta Parishad is a leading Gandhian organisation, firmly committed to peaceful pathways for achieving its objectives of land-related and other justice. In the post independence period,Bhoodan was a big initiative of Gandhians led by Vinoba Bhave for making available land to the landless. This movement continued for several years. Then there were several other struggles for land such as the 'Bodh Gaya movement', with important role of youthful Gandhians inspired by senior leader JayaPrakash Narayan. In more recent years, Ekta Parishad has maintained a continuity for efforts relating to land rights of the poor.

Ekta Parishad and its many sister organisations spread over 124 districts of 17 Indian States, constitute one of the biggest organisational network in the world which has been fighting for the land rights of the rural poor and deprived section at the grassroots. Issues relating to land reforms and equality have been emphasised very consistently by the Parishad for several decades. Even when these issues faced neglect and apathy at various levels, Ekta Parishad never lost its sense of priorities. As a result, it has emerged as perhaps the most important group involved in non-violent struggles on this issue. Its impact has become broader with growing active collaboration with about 1,000 people's organisations and movements and the widening of its horizons to include more issues of pastoral groups, nomads, fisherfolk, dam-evictees and urban poor.

Ekta Parishad's previous foot-marches in the States of Madhya Pradesh, Chattisgarh and Bihar as well as other struggles have brought justice and land rights to tens of thousands of families. Building on this strength, with the help and cooperation of several other organisations and individuals, this Gandhian organisation embarked on its famous initiative called 'Janaadesh' in 2007 when 25,000 people marched from Gwalior to Delhi to assert land rights of the poor. Specific demands included the constitution of a Land Reforms Council and enaction of a 'National Land Rights Act'.

These demands were met at least half-way by the government, but implementation and follow-up was poor.

Meanwhile, Ekta Parishad continued its efforts to forge a broader alliance on land rights issue.

Ekta Parishad's coordinater, PV Rajagopal says, 'At the same time, the assault of big mining and other projects for catering to a global market increased to such an extent that in our work area, small farmers and in particular tribal farmers, appeared to be increasingly threatened by the possibility of losing their land. The issue now was not just demanding land for the landless but also protecting the land rights of those who always had some land. We felt that the scope of our work should expand'.

During 2008-2010 about a thousand organisations and movements were contacted to form a broader alliance for land rights. At the same time, community leaders and activists from weaker sections were trained for peaceful movements and 'sataygraha'.

All these efforts culminated in a 'Jansatyagraha' in 2012. Providing the basis for its biggest effort so far, this movement said, 'If the government fails to implement land reforms effectively, the disadvantaged communities who have been denied justice and rights over land, forests and water will start a 'Jansatyagraha Samvad Yatra' on 1 October 2011. This will cover 338 districts in 24 States, culminating at Gwalior on 1 Oct 2012. After this, people will march to Delhi to get justice'.

This land march also secured the support of about 50 international land rights movements in various parts of the world.

Starting from Kanyakumari and passing through several volatile regions facing difficult land right issues in Kerala, Tamil Nadu, Andhra Pradesh, West Bengal, the North-East region, Chattisgarh, Uttar Pradesh and other States (24 States in all), the foot-marchers held many important public meetings and initiated a dialogue on several critical issues. The inter-action was particularly useful in States like Orissa and Jharkhand which were in the middle of some critical land rights struggles.

Two special strengths of the Jansatyagraha deserve special notice. Firstly, as a result of several years of dedicated efforts to contact various marginalised and deprived groups, hold public hearings and prepare reports on them, the activists of Jansatyagraha were able to incorporate a very wide range of problems and concerns relating to land struggles as well as other problem of marginalised communities. It is seldom that one sees such a consolidated effort which brings together the problems and aspirations of tribal farmers, landless dalit, farm workers, fisherfolk, dam evictees, other displaced persons (or threatened by displacement), land allottees who were not allowed to cultivate their land, urban slum dwellers and numerous other marginalised groups.

Secondly, the Jansatyagraha Samvad Yatra in 338 districts of 24 States not only made available invaluable information on the latest situation of land-grabbing, displacement, deprivation and other injustices; it also gave new hopes to many completely neglected communities who are in a serious crisis.'People were thankful that, even if ultimately we don't succeed in helping them, we at least came to listen to their woes and show our solidarity', said Rajagopal. This is why many even completely impoverished people expressed a desired to join the satyagraha's great march in October that year.

Strength of Democracy

At a time of general apathy and neglect towards such important issues, it was a very great achievement of Ekta Parishad and all the constituents of the Jansatyagraha that such a nationwide effort could take place so that new initiatives can emerge from the darkness of apathy and neglect.

It is a strength of India's democracy that such a huge, nationwide march can be mobilised by activists who remain committed to peaceful forms of struggle even in the middle of grave provocation and serious threats. Such efforts should get widespread support not just for the sake of land rights (which are very important) but also for strengthening grassroots democracy and peaceful resistance in India.

After a march of nearly one hundred thousand poorest rural people from Gwalior to Agra, an agreement was reached with the Government of India for renewal of land reforms at several levels. However, following the change of government after the 2014 elections, the agreement could not make much progress and so another struggle or Jan Andolan was launched in 2017. This also attracted a lot of attention at a time when the agenda of land reforms was getting increasingly neglected in government policy.

Ecological Movements
Linking with Livelihood Protection

> Aaj Himalaya Jagega, Krur Kulhara Bhagega (The Himalayan villages will awake today, the cruel axe will be chased away).

VILLAGERS SHOUTING THIS slogan again and again in the distant Himalayan forest of Advani (Tehri Garhwal district of Uttarakhand) were suddenly interrupted by the distant roar of approaching trucks and jeeps. 'The police has come', a youth shouted and everyone got ready to protect the trees.

When the police and the contractor's men entered the forest, these people of Henwalghati region greeted them with even louder slogans. As the contractor ordered his men to cut trees, local villagers and Chipko Movement activists rushed to hug the trees. Women were the most active. How could the axemen go ahead with tree cutting when one or more women were clasping the trees?

After making several efforts to cut trees, the contractor and the police had to withdraw. This is the now world-famous story of the Chipko Movement.

This action of the Chipko Movement in Advani forest is an outstanding example of how people's struggles to save environment and to protect their livelihood are closely linked together. In most Uttarakhand villages, forests play an important role in meeting daily needs including several types of food, fodder, fuel, leaf manure, medicines etc. while also conserving soil and water. When villagers, particularly women, encouraged by Chipko activists clearly saw the link between the ability to meet basic needs and the protection of the forests, they were quite determined to protect forests and infact, bravely faced the police both in the forests as well as at the auction site.

This incident took place about four decades ago, but the struggle was being repeated in the year 2002 in the same forest, as villagers and Chipko activists again struggled to save trees in Advani and several other areas, this time from deforestation related to the power transmission towers and high tension wires of the upcoming Tehri dam and hydro-electric project.

By this time, the fame of the Chipko Movement had spread far and wide, but such is the relentless logic of the market that the same government which sang praises of Chipko Movement at international fora was now sending its officials to cut trees (this time in much larger numbers) at one of the most celebrated sites of the Chipko Movement. The Chipko Movement may have been great and useful in bringing international fame to India's conservation efforts, but according to authorities 'it could not be allowed to stand in the way of the transmission of hydropower' to fuel the rapid spread of more industries and cities.

Rural communities leading an environment-friendly lifestyle, which has much to teach the world in terms of its harmony with nature, are increasingly threatened by the spread of big industry and giant infrastructure projects, both aimed largely at meeting the insatiable demands of consumerism. Mahatma Gandhi said, 'Nature has enough to meet the needs of all, but not the greed of even a few'. But when the 'greed' of a few becomes the predominant objective, indeed the very definition of development according to some, then nature has to be sacrificed and along with that goes the livelihood of villagers based on local natural resources.

Many struggles for protecting the livelihood based on natural resources -farmland, forests, rivers and other water sources - are also struggles to save environment. This is a source of great strength for these environmental struggles. Many struggles to save coastal ecology are also struggles to save the livelihood of fisherfolk.When the Ganga Mukti Andolan highlighted pollution of rivers by several industries, it also sought to protect the livelihood of its fisherfolk members, which was increasingly threatened by river pollution.

Although some of these struggles have achieved significant local successes at a smaller level, the task becomes more difficult when the opposition is against the larger forces tied to national or international business interests. It was easier for Chipko activists to chase away the local contractor, it is more difficult to resist incidental destruction of trees linked to a giant dam project.

The Narmada Bachao Andolan (NBA) is an outstanding example of a movement which has worked with dedication for over three decades to draw attention to and resist colossal environment damage as well as loss of sustainable livelihood of thousands of families.This struggle also clearly combined the strong urge to protect the homes, fields and livelihood of people with the protection of environment. Few movements have succeeded to the extent of NBA in documenting the threats in detail and with clarity, to the extent that even the experts appointed by donor agencies had to endorse many of these objections and concerns. In fact, several experts with deep understanding and wide knowledge of dams and river-related issues have expressed their agreement with the stand taken by NBA on many vital issues, yet the authorities were quite determined to go ahead with the Sardar Sarovar dam project as well as other controversial dam projects on the Narmada river opposed by NBA.

The Narmada valley is a place where the entire development discourse got dominated by the insistence of the authorities on creating a series of dams and so more geo-environmental changes took place in the region in the last 40 years than in the previous 4,000 years or perhaps even the previous 40,000 years.

The NBA – helped by independent experts – also worked to outline alternatives to the projects which avoid social and environmental disruption but this work has not received the attention it deserved.

Similarly, at a very early stage, the struggle against the Tehri dam project had come out with clear alternatives which avoided the massive displacement and acute hazards posed by this project, but this sound advice was ignored. Elsewhere, in the case of Koel Karo and Suvarnarekha projects, some activists who opposed these projects have been actually killed.

Despite this hostile attitude of the authorities towards these movements, they have succeeded in mobilising a lot of public opinion as well as initiating a lot of rethinking about the desirability of these projects from the perspective of sustainable and egalitarian development. It is possible that the debates initiated by these movements develop in such a way that a lot of destructive work taken up in the name of development can be stopped and replaced by alternatives which protect sustainable livelihood and environment. A lot of work remains to be done to make this a reality, and for this, dedicated professionals from many diverse fields need to work closely with these struggles.

The hostility which peaceful, broad-based and well-organised movements like NBA have faced indicates that struggles which try to defend sustainable livelihood from the onslaughts of big business interests will probably have to continue to face a lot of repression in the near future. So, efforts to broaden their base and enlist a wider support will be badly needed.

In urban areas – among prosperous sections – there isincreasing support for causes such as reducing air pollution and protecting wild life. This concern is welcome, but unfortunately, sometimes these issues are being presented in ways that threaten poor workers and forest dwellers with loss of livelihood. Environment protection measures which disrupt the livelihood of weaker sections cannot get wide support, but these can nevertheless be approved because of the support of powerful elite sections.

Such trends in which environmental issues are used to disrupt the livelihood of weaker sections should be resisted. Alternative ways of taking up environmental issues which involve the weak sections instead of displacing them should be emphasised. For example, in the context of protecting wildlife, instead of displacing villagers from national parks they can be closely involved in resisting poachers and smugglers, and trained as 'barefoot botanists and zoologists'. Protection of environment and protection of livelihood should be seen as closely related objectives, not as conflicting objectives.

Appiko Movement
Protecting Forests and Villages

THERE ARE SOME parts in the world – although these are getting increasingly rarer– where the existence of forests near villages remains the major factor in providing people a high quality of life. Several such clusters of villages can still be seen in Western Ghat region of Uttar Kannada district of Karnataka. It is true that in recent decades, the forests of this district have come under a tremendous strain due to a number of factors (including commercial exploitation for industrial-urban use and indiscriminate mining) but the destructive impact of deforestation seen elsewhere has also motivated people to get organised for the protection of remaining forests.

Thanks to this people's movement for saving forests which started in 1983, today we can visit several forests which would have vanished but for the timely action to save trees. It is a blessing to be able to stand in the middle of this greenery and to think that all this was protected by the local people. This movement for the protection of forests here is called Appiko (hug the trees). The name reminds one of the Chipko movement of Himalayan region (Chipko has the same meaning), which to some extent served as the source of inspiration for Appiko.

Just a walk in a village like Gubbigadde (people from where had been enthusiastic participants in the Appiko movement) can give a fair idea of the importance of forests in the life of people. I took only a 15-minute walk and found that there are so many types of vegetable and medicinal plants just growing wild whose uses are very well recognised by the people and they describe in detail their virtues. I was initially full of questions but soon information on the use of various plants came in such torrents that I soon grew tired of taking notes. When I entered a house, I was informed how different types of wood had been used for the upper, middle and lower portions of the door. When I expressed admiration of a small basket, my village friends hastened to show me the many different types of baskets, all made from the material obtained from the nearby forests, which were available to a single household.

Then there were the numerous types of implements, for domestic use as well as for agricultural fields – ranging from the humble broom to the sturdy plough. Above all, the forest provides the even more basic daily needs of fuel, fodder and water. The forest's gift of leaf manure has a special significance. This district is famous for its mixed gardens of arecanut, banana, black pepper, coconuts and cardamom which have been able to flourish largely because of the green manure and overall conducive environment provided by natural forests.

Due to these manifold contributions of forests it is not

difficult to understand the pain which the people of Gubbigadde (literally meaning the field of sparrows – from gubbachi and gadde) and other villages felt when the forests near their villages were auctioned to meet the needs of the plywood industry or for other industrial-urban requirements.

These villages also have a tradition of opposing anti-people forestry practices going back to the days of the British rule. The people were anxious for some action to protect trees and they listened with rapt attention when Pandurang Hegde told them inspiring real-life stories of how the people of several Himalayan villages, especially women, hugged trees to prevent them from being axed.

Around the same time, Sunderlal Bahuguna, the venerable leader of the Chipko movement from the Himalayan region, also visited Karnataka. As the people here were full of questions about deforestation and how to check this, a local youth club invited him for a public meeting in Gubbigadde village. The presence of this veteran of several difficult struggles and the inspiring stories he told about the non-violent struggles to save forests inspired the people further to initiate action for protection of forests near their own villages. People of Salkani village had already written protest letters to forest officials. Now the people were waiting for some direct action.

The opportunity came soon enough in September 1983 when the Forest Department started felling trees in the Kalase Forest. Even though this forest was located far away from the village settlements, as soon as the news reached Salkani and Gubbigadde villages, efforts were started to mobilise people for reaching the tree-felling site.

On the morning of 8th September 1983, about 160 people started their march towards Kalase, in the Kudgrod forest.

Braving rain and ignoring leaches that clung to their feet, crossing a river on a hanging rope bridge, the people rushed towards Kalase with the determination to prevent any further axing of trees. When they reached the forest, some of them rushed towards a tree which was being felled and embraced it. The woodcutters were stupefied. How could they axe a tree which was being embraced by human beings?

Surprisingly, the forest workers were not all that opposed to the basic concerns of the people and agreed to stop the work till senior officials came. On 22nd September 1983, the District Forest Officer came with scientists and influential people. At first, he said that the tree-felling was scientific and will continue, but this reasoning broke down at the felling site when a scientist accompanying the official said that the allegations of excessive damage to the forests were correct. He said that the people should be complimented for having brought this to the notice of the authorities.

Husri village was the next major site of confrontation. In 1969 a natural forest of 900 acres had been clear-felled here to raise a eucalyptus plantation. This had played havoc with the forest-dependent life of villagers, but fortunately one part of forest was still left for them. In 1983, the Forest Department sent woodcutters to fell nine trees here. The people decided to resist this. About 200 of them marched to the forest and embraced the trees. A tense situation was prevented from worsening when the official present was persuaded to take the viewpoint of villagers to senior officials and the Minister of Forests.

It will be interesting to know what happened at these Appiko confrontations in remote forests in greater detail. Pandurang Hegde has given a detailed description of the Husri movement confrontation:

About 200-250 people have gathered in forest. People in groups are near trees to prevent felling. Woodcutters (labourers) are standing helplessly. The forest officer's jeep is seen about 2 furlongs away. But nobody goes there. The jeep stops near the road. The DFO, with a Range Officer and a contractor, walks towards the people. As he reaches near the gathered people, people give respect to the officer.

Returning their greetings, he asks what are the young people doing?

One man: Hugging trees to save them from the woodcutters.

DFO: We also take care of trees. But there is a need for fuel wood. The trees cut here will be used as fuel wood in Sirsi town. If you stop felling, then there will be fuel wood shortage in Sirsi and grave problems.

Villager: You can take dead and dry trees but not green ones.

Another person: Once you cut down this forest there will be no trees left. What will you do once the total forest is gone? From where will you bring fuel wood?

Villager (An old man): Yes? You plant teak and eucalyptus (moving his hand towards eucalyptus plantations). And is that you are going to plant? And will this be used for fire wood?

DFO (a little puzzled): That was done previously. But here, we will plant trees which will yield firewood.

Villager (a middle-aged man): This is good. But where will we get green leaves for manure and wood for agricultural implements?

Another man: We cannot use eucalyptus and teak for green manure nor for manufacturing agricultural implements.

DFO (does not answer these questions). You see there is urgent need for fuel wood. Allow us to cut trees, otherwise problems will be created in town. Dead and dry trees as you suggested is good idea. But it takes time. So, allow us at least a year's time.

Villager: If we allow you a year, our only forest near the village will be cut down. Then? Where are we to go? Can you get us back these forests? The trees, which are useful for us? Forests are not just for villagers. There are people in other areas too. They also need fuel wood, timber. As you get grains from outside, we also have to give forest products.

Young man speaks: Sir, when we have enough then whatever is left, we will give to others. If we are hungry can we give out food to others?

DFO (no answer, diverts the topic): Look, if you want to solve these problems why don't you go and meet the minister?

The elderly person: Sir, as you came here in the midst of forest and saw the situation, we can make you understand our problem, what use is it to go to a minister in Bangalore? He is too busy to understand. And, moreover, he will not know any of our problems without visiting this place. We have already sent memorandum. Like you have come, minister should also come here to understand the situation.

DFO: The order to fell trees has been signed by the governor. I cannot do anything. I cannot order to stop cutting of trees.

Villager: We agree you cannot change the order. But we will not allow a single tree to be harmed by any woodcutter. We will hug the trees.

Another person:We want forests, which give us rain, manure, fuel wood for agricultural implements. We will not allow it to be used for fuel wood and timber. Forest is ours; we need it, it is for our survival. They produce soil, water and fertilisers.

As the forest officer tries to go away, a part of the crowd discusses about gheraoing him for the whole day. However, the organizers of the movement persuade them and make them understand that this is different from any political movement. Being surrounded or gherao is an expressing anger. 'It does not lead us anywhere. Our goal is to change forest policy. So, it is of no use harassing any individual.'

The DFO goes away. The villagers hold a meeting and take a pledge to protect the forest at any cost, even at the cost of their lives.

In the first phase, the Appiko movement had spread within a short time to 8 areas – Mathghatta, Salkani, Balegadde, Husri, Nedgod, Kelgin Jaddi, Vanalli and Andagi.

In the last week of December 1983, the much-awaited visit of the Forest Minister to Kalase and other affected areas came. People turned out in large numbers to present their viewpoint to him. The Minister gave specific orders to stop the felling of several marked trees and said that in future, only dead and dry trees will be cut from here.

In April 1984, some Appiko activists decided to go on a long foot march to take their message to a wider area. This march started from Sirsi town on 10th April 1984 and after covering about 650 kms, the marchers returned to Sirsi on 29th April 1984. Such foot marches enabled the movement's message to spread to a wider area and also helped the activists to get a firmer grasp of the realities.

Regeneration and Greening

Within three years, the movement had also spread to the districts of Shimoga and South Kanara districts. In some villages, resources were mobilised initially be daily collection of handfuls of grains.

The traditional theatre of Karnataka 'Yakshagana' was adopted to spread the movement's message. The Appiko movement was not concerned only with confrontations. Following the ban on green felling after 1989, the movement gave more attention to regeneration of degraded land.

Pandurang Hegde and other Appiko activists started getting frantic calls from other areas to come to help in forest protection initiatives. This was different from earlier efforts as now they had to go to new areas. Here they did not have previous familiarity with the people with whom they were interacting. In some places, local politicians tried to make use of the popularity of the new movement to serve their own narrow interests. All this sometimes created a difficult situation for Appiko activists from Uttar Kannada. They had to visit the site of tree-felling and then mobilize people to check this. In Coorg, they learnt that the lawyer who contacted them was based in Bengalaru and came here only from time to time. So, the work for them increased.

Undeterred by these difficulties Pandurang Hegde and Appiko activists plunged themselves into spreading the movement in many new areas. As they found the right contacts, several reliable and committed local allies emerged who took on a lot of responsibilities. These included AC Subbaiah in Coorg and Shampa Daitota and Hari Rao Marlaje in South Canara.

Several early successes became possible because of the high levels of dedication and commitment shown by several emerging village leaders and activists such as Mahabaleshwar Hegde and Vijaya. They came forward not only to hug trees in very difficult conditions, but also contributed with their special skills in music and songs. Others participated in street plays and folk theatre (Yakshagana) presentations of the issue to take the message of the new movement far and wide. After her marriage, Vijaya settled in Bengaluru and helped to spread the Appiko message there. GG Hegde came forward to offer local support to Pandurang Hegde in such a way that he could devote all his time and energy to spread the new movement. In the success of the Belagadde movement, initial groundwork had been done by two youths Ishanna and Mablu. There was excellent media coverage by the Kannada and English media, and this helped people in several distant places of the region to initiate tree protection work in their own ways. They needed the help and support of Pandurang Hegde and other Appiko pioneers to take forward their new forest protection efforts.

Renewable Energy
Decentralised Rural Model

THE MANY-SIDED BENEFITS of renewable energy sources are now widely accepted. The challenge now is to create a model of renewable energy development that is most suited to the needs and potential of our country.

A decentralised, rural-based model of renewable energy will be most beneficial for a country like India. Nearly 70 per cent of the population lives in villages but the over centralised energy model that exists today has been found inappropriate in many ways for meeting the energy needs of the rural population. A decentralised renewable energy model, on the other hand, is particularly suitable for meeting the needs of villages, particularly the more remote and inaccessible villages which are least covered by the centralised grids.

While planning such a decentralised energy model, it'll be most useful to link up this planning with the panchayat raj institutions in the country (PRIs or rural local self-government institutions). The functioning of panchayats at present is far from ideal and there is growing agreement that PRIs should be reformed in such a way that the gram sabha or the assembly of all villagers can play a more effective role. Village plans should be prepared on the basis of extensive discussion in the gram sabha with adequate opportunities to weaker sections and women to present their views. The planning for renewable energy should also be linked to this process.

For example, there is a lot of potential for the development of watermills in hilly areas. While travelling in many hill villages, I noticed that traditionally water mills played a very useful role, particularly for milling wheat, but there has been a decline in recent years. Several well-informed persons told me that still, watermills can be very useful and the generation of very small-scale hydel power (micro hydel) can be added to watermills. There will be no adverse effect while a lot of employment will be generated. This decentralised model is very different from the centralised model of inflicting large dams on villagers with highly-disruptive social and environmental impacts in many cases.

When thousands of gram sabhas prepare plans for watermill-cum-micro hydel, keeping in view local conditions and sensitivities so as to minimise the possibility of any adverse impact, then opportunities of making available electricity to villages at a low cost and with minimum adverse effects are opened up.

While each effort is small, but adding up the efforts of thousands of panchayats, this will lead to considerable electricity generation in a safe and environmentally secure way.

This is precisely the beauty of the decentralised model that it makes this possible to harness the potential of so many local people for obtaining best development results.

In this context, a very important contribution has been made by the Barefoot College (BC), based in Tilonia (Ajmer district, Rajasthan). BC has spread solar energy in very remote villages and what is most remarkable is that this task was mainly performed by villagers (with up to class ten school education). Barefoot College has been organising many trainings sessions in solar energy production. After about six months training, many villagers emerge as 'barefoot solar engineers' although on-the-job learning continues long after this.

The work done by BC and the success of its barefoot solar engineers has increased the possibilities of harnessing the neglected talent of villagers and providing them a lot of employment in decentralised renewable energy systems. If we did not have this example of Tilonia's barefoot solar engineers before us, then probably local villagers would not even be considered for many technical tasks which they can certainly accomplish with good training. Success of 'barefoot solar engineers' can be replicated in the case of several other renewable energy sources as well.

The model, in sharp contrast to the Tilonia model, is one in which a company installs a solar system in a village without involving the villagers in the task. This model neither increases the self-reliance of villagers, nor their skills or employment opportunities. The Tilonia model, which increases self-reliance, skills and employment prospects of villagers should certainly be the preferred model.

Recently the cause of decentralised renewable energy-based systems has been helped by the campaigns of Greenpeace. In particular, attention has been drawn to an intensive campaign in Bihar which convinced many people how useful decentralised renewable alternatives are for villagers who have been treated shabbily by the centralised electricity systems and grids. As this organisation's campaign asserted strongly, 'Greenpeace believes that a resurgent Bihar can chart an alternative development pathway via decentralised energy infrastructure to provide for the energy needs of the rural population in an equitable and sustainable manner. Bihar needs an energy revolution now and decentralised renewable energy can fuel that change'.

Greenpeace has demanded that a State based regulatory framework should be developed to encourage utilisation of renewable energy through a Renewable Energy Law. The new law should be drafted with proper and wide consultations. Adequate infrastructure should be created at the government level for successful implementation of decentralised renewable energy.

These and other various efforts for the progress of decentralised renewable energy systems should link up with thepanchayat raj and reform of panchayat raj should take place so that the chances of such a decentralsied renewable energy model succeeding in more and more villages can increase.

Movement for Transparency
Struggles of MKSS

THE POST-INDEPENDENCE PERIOD has seen a fast spread of voluntary organisations in India. While some of them have done very good work, even outstanding work which has opened up new paths, a widespread feeling has been that on the whole quantitative growth has not been matched by qualitative aspects. One manifestation of this has been that issues relating to ethics, transparency and code of conduct, even when raised from within these organisations, have not received the necessary attention. In fact, some of those who persisted in raising these concerns had to face a lot of opposition from within these organisations.

A conviction started growing among some activists to explore alternative forms of social mobilisation and organisation. Another context was provided by the need to come out of the limits imposed by project-oriented work, so as to link up more closely with the wider social and political change.

During the late eighties, three activists in Central Rajasthan were thinking along somewhat similar lines, although they had other concerns too. They knew for certain that they were not going to accept any institutional funds, and that they were going to live among the rural poor more or less according to the norms of the people. They would be accepting only the legal minimum wage rate for workers. This money, they hoped, to be able to raise from lectures, writings, small donations from friends etc.

These activists were also clear that they were not going to start a political party. The path they chose for themselves was broadly that of a 'non-party political process'. Although how exactly this will evolve was to be still seen in future with its several uncertainties, but this much was clear that this entire journey would be guided by several non-negotiable values such as commitment to the needs and aspirations of the poorest and weakest and rejection of discrimination and injustice based on caste, gender, race and religion.

Although these three activists thought for some time about moving to Jhabua district of Madhya Pradesh, their final decision was to continue to remain in Rajasthan where two of them already had experience of working for several years.

Only one of these activists was already well-known to the wider world. Aruna Roy's decision to resign from her Indian Administrative Services job (her resignation was accepted in 1975) had earlier attracted a lot of attention. Keeping in view her firm commitments to social justice, there were high expectations that she woud use her position as a highly placed official to push these concerns as much as possible. But Aruna chose to come out of the limitations of this system and devote her life more directly and explicitly to social concerns by working with the Social Work and Research Centre (SWRC), a voluntary organisation founded by her husband Bunker (Sanjit) Roy. Also known as the Barefoot College, the SWRC had initiated pioneering work in several areas and also raised wider issues relating to code of conduct within the voluntary sector.

Aruna made a very valuable contribution here for about eight years (1975-83), particularly in the mobilisation of (non-funded) women's groups in villages which remain a very strong part of the many-sided activities of SWRC. At a wider level, Aruna was increasingly involved in women's activities and other kid of mobilisation particularly in Rajasthan (1983-87).

At the mid-career age of 40, most people like to build on past strengths instead of venturing to entirely new areas. The reluctance is likely to be even greater, when the new area is extremely uncertain and difficult. But Aruna Roy took this decision in 1986-87 to explore the path of a 'non-party political organisation'.

The second member of this team of three was Shankar (age 32 years in 1986-87). For Shankar too, this was a difficult decision as after several years of moving from one uncertain, difficult job to another (in fact, he recalls changing 17 jobs) he had just begun to settle down with some years of very creative work at the SWRC. It was here that his hitherto hidden genius as a communicator flowered in diverse art-forms including puppets, songs and street-plays. Shankar acquired a reputation – which has continued to this date – of a communicator who talked about very serious issues very skilfully, yet never missed a chance to spread joy and laughter among people and colleagues.

The more likely path for Shankar would have been to

continue his stable job which was providing him enough creative outlets for his talents. The fact that he had a family to support made a big and uncertain change all the more difficult for Shankar. But ultimately, he convinced his wife Anshi, of the need for this change and the couple with their three children started preparing themselves for the more uncertain life ahead.

The youngest member of this team was Nikhil Dey, who had just turned 24 years in 1987. As a teenager he had always yearned to be a part of a wider social and political process of change. Even when he went to the USA for higher studies, his heart remained in this and he returned sooner than expected. Initial exposure to the political party process was disappointing for this young man of high ideals, and the conviction of working at the grassroots with workers and peasants became stronger day by day. It was with this idea that he went to several parts of the country and explored several options. He could barely speak a few words of Hindi initially but learned quickly. Ultimately things worked out with Aruna and Shankar and after several discussions the team of three, along with Anshi and three children, settled down in a hut of Dev Dungri village in Bhim sub-division of Udaipur district (now this village falls under Rajsamand district).

At that time, young Nikhil could not have imagined that within the next two decades along with his colleagues he'll be able to make extremely important national level contributions, and so many youths will feel inspired by his example.

Aruna, Shankar and Nikhil had fairly wide age-difference and also different temperaments, but unity of purpose proved much stronger and they have been together since then through a succession of difficult challenges and complex problems as well as great achievements.

The choice of Dev Dungri was guided mostly by the fact that Shankar's sister had a spare hut here (her husband's family had settled in a different village) which could be used by these activists without incurring any expense. This was an area frequently exposed to droughts which meant that most rural families here badly needed drought-relief work or rural employment schemes. Most villages had OBCs (mostly Rawats) as the most predominant community, followed by dalits and other castes. Due to frequent drought periods peasants were also reduced to workers in most cases – whether drought-relief workers or migrant workers is another matter.

Trying to understand the social-economic realities, the newly arrived social activists in Dev Dungri were initially in no hurry for quick actions. In this phase Aruna and Nikhil also found time to teach children at their home. However, these were difficult times for villagers and they themselves started coming to them with suggestions of what needed to be done.

However, it was only in 1990 that the Mazdoor Kisan Shakti Sangathna was formally launched on May Day.

The three activists and their handful of initial supporters at that time were certainly keen to take up many good initiatives that the villagers of this area badly needed, but they could not have imagined what they would be initiating soon (right to information campaign, social audits) will have such a great impact at the national level. This has turned out to be a classic case of a small beginning that achieved great things.

The right to information campaign and social audits grew so organically out of the initial struggles of the MKSS for the rights of poorest sections that the poor and illiterate people of the area could recognise the true worth of RTI (right to information) and social audits better than even urban educated people. This was the true strength of the RTI movement in the MKSS region. It was a great achievement of MKSS that it could take forward this initiative to a national level and played a major role in getting some of the best talents at very senior levels (for example some of the senior journalists and the Press Council, senior academics and retired judges) together for a national campaign (NCPRI).

At a later stage, when some persons tried to misrepresent the situation and tried to take almost all the credit for the RTI law in India, mistakenly presenting some drafting (making some changes and additions in the several drafts that had already been prepared) as the entire RTI campaign, the MKSS did not even feel the need to challenge this misrepresentation, maintaining a dignified silence as the reality of the entire RTI process and movement in India is only too well known to the persons who were a part of this.

Later, the MKSS also played an important part in the campaign for rural employment guarantee, as also in better implementation of this law. On the other hand, not so well known but no less valuable is the role the MKSS played in some local issues of Rajasthan such as a successful campaign (taken up in collaboration with other grassroots organisations) against the takeover of lakhs of acres of village commons for biofuel plantations and related measures.

The contribution of the MKSS, working with NCPRI, in a reappraisal of the Jan Lokpal draft and presenting an alternative basket of anti-corruption measures was timely and valuable. It opened up new dimensions to a narrow, hurried debate driven more by rhetoric than reality.

Nearly three decades after the informal beginning of this initiative, the MKSS can take legitimate pride in its many-sided and invaluable achievements. The achievements appear even bigger if we look at the admirable record of several youths who were associated with the MKSS for some time and then left to work in other areas, maintaining close links and also carrying with them rich experiences of their association with the MKSS.

Doctors Devoted to Serving the Poor

Some Dreams do become a Reality

ABOUT THREE DECADES ago, a small group of bright young medical students at the All India Institute of Medical Sciences in Delhi shared a beautiful dream. One day, they told and re-assured each other, 'we'll all pool our skills and energies to start a health programme in villages that will give genuine health care to the rural poor. We'll start a hospital from where no patient will have to return due to lack of the capacity to pay for medicare and medicines. We'll do all we can to take medicare to the poorest of the poor.'

These students discussed this dream again and again and they shared it with some of their friends in other leading medical institutes such as CMC Vellore. As the years passed, they also shared the dream with those friends whom they hoped to marry. When the dream was shared at this level also, the resolve to translate the dream into reality was strengthened further.

These days not many young people dream about serving the poor. Not many dare to dream about giving up highly-paid secure city jobs to choose an uncertain difficult future in remote, unfamiliar rural areas. And even among the few who dare to see such dreams and sustain them for several years, the chances of realising these dreams are frankly not very high because overall, the conditions are rather hostile for such efforts.

But fortunately in the present case, the determination and dedication of our young doctors has led to the actual realisation of their dream to such an extent that about 1,000 patients drawn from about 2,000 remote villages (mostly poor people in a radius of around 200 kms.) are treated every week. The ideal of not turning away any patient has become a reality to a very considerable extent. Above all, these doctors have been able to take the lesson of their work in remote rural areas to policy-making level so that a potential for policy changes has been created which can benefit millions of people in the near future. And all this has been achieved in the middle of various adversities and difficulties at almost every step.

Go back in time to 13 years when the first three doctors have just come to Bilaspur town in Chhattisgarh and are trying to use an old, long-abandoned building (in Ganiyari village) as a clinic after some renovation work arranged by their friends. One of them is 7-months pregnant. The other two doctors (Yogesh and Anurag) ride a rickety scooter to start their first OPD day. They are young, they are excited, they are just too eager to start serving the poor. But the morning turns into scorching hot afternoon, then slowly the sun starts receding and darkness sets in. Not a single patient has come yet. Silently, the two doctors leave. Next day exactly the same experience is repeated. Two long days, zero patients.

Then on the third day, the first patient comes in. Next day 4, then 7 then... And sooner than they could adjust... they were neck deep in work. This has been the experience ever since then. Doctors here are so overwhelmed with patients that despite their best efforts, there is a never-ending long que.

We're talking here about Jan Swasthya Sahyog (JSS), one of the most talked-about pioneering initiatives of young doctors in the country to make a real breakthrough in improving rural medicare. This effort was initiated by a handful of doctors as its founder members – Yogesh Jain (Secretary), Raman Kataria, Anurag Bhargava, Biswaroop Chatterji, Sathyamala, Rachna Jain, Anju Kataria, Madhuri Chatterjee, Madhavi Bhargava, Pramod Upadhya (who played a particularly significant role in appropriate technology programme) and BR Chatterjee.

Early Days

> We had looked at so many places to start our work that we were getting exhausted. Our main criteria was that we should be able to reach the poorest of the poor, the people most in need. Also if some support and contacts existed in the form of voluntary work already being done here, than that would be helpful. These conditions were substantially met in the present project area which we finally selected with the help of local friends like Praful Chandel (who still plays a very useful role in JSS).
>
> Our aim was to understand the needs of villagers and then meet these needs effectively. So initially, we spent a lot of time in visiting villages and talking to people. Here, our colleague Dr Sathyamala, in particular played a very helpful role in helping to form a proper understanding of the needs of people. Yes, language was a barrier to begin with but when there is such an intense desire to understand eachother, then such barriers can be overcome.
>
> —Dr Raman Kataria, Founder-member, JSS

This has been a rare effort in India (actually there should/could have been many more), when a group of doctors got together to unitedly create a community health effort. Some doctors joined at a later date while some took a break for some time and have returned or hope to return after some time. On the whole, this has been a remarkable story of several doctors sustaining united effort (overcoming some disagreements for the larger cause) with such relentless hard work, impeccable integrity and great dedication that it could become a role model and source of inspiration within 13 years, in the process also treating/serving over 2 lakh patients drawn from about 2,000 villages in 8 districts of Chattisgarh and Madhya Pradesh.

JSS works at a three-tier level of community health workers in hamlets, health sub-centres and the main hospital at Ganiyari (Bilaspur district). One of its main priorities is a community health programme, providing preventive and curative services, based on about 120 trained village health workers, elected by their respective villages, in about 60 villages of Kota and Lormi blocks mainly (but not only) inhabited by tribal communities (mainly Baiga and Gond).

Clinical services provided by the JSS include an outpatient clinic at Ganiyari, well equipped with a dispensary, radiology and laboratory services, that has provided extremely low cost and effective care through about 5 lakh consultations. Patients come here from over 200 kms. radius area inhabited by about a million people. This hospital has a well-established, well-deserved reputation (recognised also by senior health officials) for good-quality comprehensive medical, surgical, paediatric and obstetric care in the region.

This main hospital includes an inpatient ward with 50 beds and an operation theater complex (including 3 major operation theaters and a labour room) that has provided high-quality surgical services to above 20,000 patients.

Emergency!

A 3-months old child was brought to Ganiyari hospital with a very swollen stomach, result of intestines obstruction. The child is an advanced stage of toxic megacolon. Both the parents are TB patients. There seems little hope, but prompt and careful treatment including surgery revive hope. Other surgeries follow after some gap and the child survives to lead a normal life.

A 27-year woman comes in great pain because of sceptic abortion. The quack who performed the abortion with a stick laced with some herbs had damaged her uterus, ureter and the vessel. Ganiyari hospital was still in its early days and a lot of running around had to be done to arrange the basic pre-conditions. Some hospital staff donated blood and surgery continued for 7 hours. Finally, the woman's life was saved.

A one-day old child of a poor family who has abnormally formed feeding and respiratory passages, is referred by a medical college to Ganiyari hospital. Here the child gets prompt care, including surgery and survives.

Three subcentres in three forest village clusters are located at Shivtarai, Semaria and Bamhni villages. These subcentres together serve over 150 forest and forest fringe villages. Each sub centre is served by two senior health workers and supported by doctor-based outreach clinics every week. Doctors and paramedics have to walk across rivers in spate which completely drench them on their way to Bamhni sub centre.

JSS also has an appropriate technology centre for development, adaptation and validation of low-cost health care and diagnostic technology suitable for application in rural areas.

JSS conducts operational, observational and action-based research on priority public health problems with special emphasis on childhood illness, tuberculosis, malaria and leprosy.

JSS has been organising training programmes not only for its own health workers but also health workers of other voluntary organisations and people's movements as well as the government's health personnel. Also, it has been providing advice and inputs as a Technical Resource Group for the State government (Chattisgarh), the Planning Commission and other agencies.

Research and training efforts of the JSS have linked it to possibilities of much wider improvement in health care both at the level of policy and implementation, in the government sector as well as in the voluntary sector. This wider reach of JSS is further helped by its advocacy efforts which have a special emphassis on meeting the health and nutrition needs of the poor. A particularly effective part of this advocacy has been the advocacy for a national policy for medicines, based on need and not on greed. The JSS has been involved in public interest litigation on this issue which has the potential of leading to important policy level changes for making available low -cost and rational medicines to people.

JSS has made presentations on this issue before Parliamentarians' group, Special Parliamentary Committee as well as various other important fora. Presentation on other policy issues have been also made before various professional bodies such as the Indian Cardiac Society, the ICMR and the Indian Academy of Paediatrics.

It is a reflection of the broader approach of the JSS to link health to nutrition issues that it has initiated a programme of rural creches called *phulwari* which provide care and nutrition to children in about 90 rural hamlets. Similarly, JSS is one of the few health initiatives which also has a programme for improving agriculture yields using the agro-ecology approach, organic methods and indigenous varieties.

One basic thinking common to all these activities of the JSS is that inequality in health as well as broader inequalities have to be challenged, fought and reduced significantly so that the poorest people can have access to good healthcare as well as nutrition. Another basic priority has been to establish a relationship with patients which is based on the values of humanity, dignity and compassion.

These special strengths of JSS have started attracting a younger generation of doctors, several of whom (particularly from Maharashtra) are now contributing to the further programmes of JSS in meaningful ways. One hopes that with the passage of time this inspirational role of the JSS will grow, and many young doctors who learn these basics here will go on to create JSS-type institutions in several parts of the country.

The statistics given are for the time of the writer's visit.

Alternative Politics
Mobilising People for Protective Role

BY POLITICAL PROCESS or politics we mean the mobilisation of people to obtain their support for certain ideas, actions, policies or programmes. This mobilisation may or may not include the aim of capturing state power or establishing a government based on these ideas and policies.

The term alternative politics has come in vogue to denote the mobilisation that takes place on ideas and policies that differ radically from dominant trends promoted by most governments and powerful corporate interests.

There is a great need to build up people's mobilisation on the basic protective role of human beings on earth. Among millions of life-forms on this planet, human beings alone have the capacity to work in a planned way for the protection and wellbeing of all forms of life as well as the protection of the environmental conditions and habitats which sustain these numerous life-forms. This unique capacity of human beings defines their main role on earth which is to protect all forms of life.

The creation of a world in which most people recognise the protection of all forms of life to be the main aim of their life is a very noble endeavour but it is also a huge, difficult task. People who have the courage of their conviction first bring their own life in conformity with this objective – not an easy task in the existing situation – and then try to mobilise more and more people for such a change. Several organisations and movements are already working with somewhat similar (although perhaps not quite the same) aims. They can become allies in this task. As more and more people are convinced about the significance and centrality of the protective role of humanity in creating a better world, they can form small groups to share larger responsibilities. These groups can be linked to each other.

A significant task of these groups is to evolve a detailed framework of the world in keeping with the protective role of humanity. It is very well to emphasise that the basic role of human beings on earth is to protect all forms of life, but further it has to be worked out what this basic principle means in the context of the most pressing questions, problems and issues of our times.

As these issues are discussed in detail, many different points of view will emerge, and it'll be important to evolve a consensus without sacrificing the basic principle. For example, once it is accepted that the greatest role of humanity is to protect all forms of life, then one of the basics that emerges from this principle for our practical daily life is that our food should be vegetarian. However, this may not be acceptable to many people for various reasons. For people of some regions, for example very cold areas, it may be difficult to survive without some use of meat. So, a consensus may be reached to exclude such areas from the requirement of vegetarianism, but still to emphasise the principle of as less pain to animals as possible. For some areas, such as coastal areas, eating and catching fish may be such an integral part of food as well as livelihood for some communities that it is very difficult to visualise their life without catching and consuming fish. In such cases catching and eating fish can be permitted while emphasising sustainable practices (so that any fish species are not threatened) and at the same time taking care to protect fish and other aquatic life from all avoidable suffering. Other communities which for some reason cannot adopt vegetarianism can also be asked to reduce the pain and suffering of (to be slaughtered) animals as much as possible.

This example reveals clearly the difference of two approaches. One approach remains quite rigid and in the process several people are excluded. The other approach allows for some flexibility so that more people can be brought closer to the basic principle. If we go by the first approach, then a lot of people may remain alienated from the groups who spread the message of protective role of humanity. It may be better to provide them an opportunity to come closer to the principle within the limits of their own constraints relating to their culture, tradition, habits and livelihood. Moreover, once a person accepts the basic principle of minimising the pain and suffering of all forms of life, it is quite possible that over a period of time he or she may fully adopt vegetarianism, initiating a trend within his own community which may be more influential than any message coming from outside sources.

A related key concept is that of simplicity and frugality.

While the dominant trend in present-day consumerist world is to acquire/use more and more goods and services, the principle of simplicity/frugality encourages entirely different thinking. Here we learn to ask - what are the things I can do without? What are the things that are truly essential for my health, well-being and work? I should have only this much and no more. In keeping with this thinking, our need for money is also reduced to meeting real, basic needs and some saving for emergency situations. Our own needs restrained in this way, we have more time, resources and efforts for the welfare and protection of others.

As the mobilisation of people proceeds an important issue that needs to be resolved is whether this mobilisation should also start moving in the direction of contesting elections and striving to capture state power by peaceful, democratic methods, mainly contesting elections.

One reason why some people oppose such an effort is that present day electoral politics involve a lot of corruption and a dominant role by the richest persons and corporations. They first give money to political parties and politicians to fight elections and later get even bigger favours from them.

So, the alternative political process will have to take the greatest care to keep away these corrupting influences. Firstly, this political party should function in a completely transparent way. Detailed records of all income and expenditure should be maintained and this should be open to examination by any citizen. All expenses for an election should be raised within the constituency of this election by local people. So the political party does not have the responsibility or burden of election expenses at all. It mainly provides guidance on matters of policies and principles.

The second problem with existing political parties in our electoral system is that they tend to orient all their rhetoric and activities to the predominant aim of somehow winning election and capturing power. However, the alternate political process will not compromise on any important principle. It'll take its programme among people as best as possible and if people are not yet ready for this programme then it'll be better to wait till they are ready to accept it and vote for it. Till that time the alternative political party will be happy to survive as an opposition party, if necessary, even a small opposition party, without giving up the hope of one day spreading its support base among most people.

GANDHIAN ACTIVISTS AND SCIENTISTS

The ones who are crazy enough to think they can change the world are the ones who do.

—Mahatma Gandhi

First they ignore you, then they laugh at you, then they fight you, then you win.

—Mahatma Gandhi

An ounce of practice is worth more than tons of preaching.

—Mahatma Gandhi

Kunwar Prasun

Fearless Activist Gandhian Thinker

Kunwar Prasun was like an earthern lamp which lights the path of others, and this light will continue to illuminate the path of activists long after his death.

—Sundarlal Bahuguna

IF ONE IS looking for the example of a person who remained a tireless Gandhian activist all his life, without seeking any reward or fame, then the name of Kunwar Prasun comes to mind very quickly.

If only one identity is to be emphasised, then Kunwar Prasun should perhaps be remembered as a leading activist of the *Chipko* (save the trees) Movement because of the very prominent role he played in this movement in Henwalghati region of Uttarakhand, along with his close colleagues like Dhum Singh Negi, Vijay Jardhari, Sudesha Devi, Ghanshyam

Sailani and others, all of them guided by veteran Gandhian leaders Sundarlal Bahuguna and Vimla Bahuguna. In the case of Prasun, emphasising just one aspect can never be adequate.

If one looks only at the activist side of Kunwar Prasun, then he started as an activist of the anti-liquor movement when he was still a school student. Then he went on to play a leading role in the movements to prevent destructive mining practices and to oppose construction of large dams. He was not from a dalit family but he contributed richly to dalit activism.

At a very young age he became a strong proponent of 'equal education for all' and organised a long foot march on this theme. He was a leading activist of the student/youth organisation called Chatra Yuva Sangharsh Vahini and became its coordinator for the State of Uttar Pradesh (this was before the formation of a separate state of Uttarakhand). He was also a consistent campaigner against religious divide, communalism and related ideologies. While he was a champion of national integration, his thinking went beyond this and he was truly a world citizen. His last days were devoted to conserving traditional seeds and popularising organic farming. He emerged as a very strong advocate of agro-ecology and farming practices which are in harmony with environment protection. Along with Vijay Jardhari, he played a very important role in the movement to save traditional seeds (the Beej Bachao Aandolan).

Kunwar Prasun was also a very capable writer and journalist who had a style distinctly his own. He was one of the best examples of what may be called a barefoot journalist. With minimum of support from anywhere, he travelled to some of the remotest villages and reported from places from where perhaps no journalist had reported earlier. He had a distinct style which is best described as 'telling the truth as it is'. As his co-writer on many occasions, I worried endlessly about the impact of writing too openly about some sensitive issues (for example the practice of prostitution in some areas). Prasun never questioned my apprehensions, but promptly went on to write as he had always done!

Unlike many other Gandhian writers, Kunwar Prasun took delight in taking digs at those persons about whom he was convinced that they were guilty of various malpractices. Knowing his precarious finances and the needs of his family, I worried about these powerful persons (who were at the receiving end of his sharp pen) taking legal (or other) action against him. My fears were not unfounded and Prasun got several legal notices. But just as my pleading did not change his writing style, all these legal notices did not affect his writing style either!

While Prasun's reporting and comments on local issues were widely published, I regret deeply that his views on national and international issues, particularly in terms of evolving alternatives to the existing system did not get adequate attention.

After Prasun's untimely death in 2006, when I sat down with my wife Madhu to compile a selection of his writings, we were amazed at the depth and diversity of his writings. If these had appeared in English, these would have surely attracted international attention. Prasun (along with his writer colleague Navin Nautiyal), was very involved in whatever he wrote about, and he never tried to hide his likes and dislikes. For some time, he reported for one of the largest Hindi newspapers published from Delhi, but then resigned to work in a free-lance capacity.

Prasun also wrote in Garhwali and used the folk-forms (like Mandaan) brilliantly. His Garhwali talks and stories were broadcast regularly from Najibabad radio station.

Prasun was a thinker in the Gandhian tradition, deeply committed to peace and non-violence, but he often went beyond what is generally written by Gandhiji's followers, because of his more intense inter-actions with the society he wanted to change and also because of his pronounced tendency to speak out openly whatever he felt. Unlike the Gandhians who are tied to various institutions, Prasun was always very frank and even blunt in his opinions. Views of such a person on national and international issues would have been very interesting and useful, but due to various reasons he could not do such writing very frequently, and even less was published.

Prasun held his convictions with courage and was never afraid to be isolated. In fact, on quite a few occasions I've seen him speaking in a setting of 'Prasun vs all the others'!

At a time when it was becoming almost a fashion to negate some basic ideas of Gandhiji, Prasun spread Gandhiji's ideas with a passion very rarely seen these days among Gandhiji's disciples.

Prasun was born in a high-caste farmer family of Tehri Garhwal district (Uttrakhand). In his youth, he deliberately used a name which concealed his caste identity, as this was the practice of some activists who were attracted to JP's movement (the movement led by veteran Gandhian Jayaprakash Narayan) and his call for creating a casteless society. In fact, so intense was Prasun's advocacy for dalit rights that some of his colleagues in various movements thought that he belonged to a dalit family!

Once when he was in college, Prasun and his classmate were asked to get a certificate attested by an 'eminent' person. Prasun promptly went to a nearby cobbler (a dalit) and got his signature on his document. When the teacher yelled at what he had done, Prasun replied calmly that in his opinion the cobbler was an 'eminent' person.

As I stated earlier, Prasun participated in the anti-liquor movement even when he was in school. The teacher who inspired him for this is Dhum Singh Negi, who later became his co-activist in Chipko and other movements, although Prasun always continued to call him Guru (or teacher). Prasun's first movement succeeded in driving out a liquor-seller from the village.

In college, Prasun came in contact (and was much influenced by) the leading Gandhian activist of Garhwal named Sunderlal Bahuguna. This also proved to be almost a life-long association. Prasun also had a close relationship with Sunderlalji's wife, Vimla Bahuguna, who became a mother-like figure for several young activists like Prasun. In college in Tehri, Prasun also met Vijay Jardhari, who also remained his co-activist in several movements.

Those were the days of an intense anti-alcohol in Uttarakhand and Prasun was a very active participant, smashing liquor bottles at one place or confronting the liquor contractor at another place.

After completing his college, Prasun worked as a school teacher for a very short time. At an early stage he also got involved with the movement for social and political change led by JP and was lucky to escape when the government imposed emergency and cracked down on this movement.

From 1977 onwards, Prasun was intensely involved in the Chipko movement in Henwalghati region of Tehri Garhwal district. When trees were being auctioned by the government, he went to villages to mobilise people against tree-felling. Such were the skills of this young activist that in a few days he was able to prepare villagers, particularly women, to oppose the felling of trees even though earlier the contractor had used his money and agreement with the government to get influential villagers on his side. Prasun was not just a fearless activist, staying in forests on cold desolate nights to protect trees, he was also very determined and persuasive and had great skills in communicating with villagers.

Prasun's another special skill his co-activists remember is that of finding very effective slogans. Some of the slogans he coined became world-famous.

Later he participated in the protest against tree auction at Narendranagar when he was arrested along with several other villagers. Undaunted, he went to other places like Badiyargad and Khuret to participate in the Chipko movements there. Khuret proved specially lucky for him where he met and married Ranjana, whose tremendous support enabled Prasun to contribute to several more movements.

For sometime, Prasun became deeply involved in the anti-Tehri dam movement. Then he was also in the thick of the movements against destructive mining practices in Nahin Kala and Henwalghati.

In the last phase of his life,Prasun was deeply involved in conservation of traditional seeds and in the spread of organic farming practices.

Simultaneously he opposed the so-called 'green revolution' farming with its emphasis on chemical fertilisers and pesticides as well as corporate agriculture. He was very firmly opposed to GM crops and patenting of seeds or life-forms. He opposed all efforts to replace traditional cropping patterns and rotations with new commercial crops.

Together with other leading activists of the Beej Bachao Andolan, like Vijay Jardhari, he went (as a part of foot-marches or in smaller groups), to remote villages to gather information about traditional crop varieties and collect seeds where these are still available. We were able to publish his notes on documentation of over 328 traditional varieties of paddy grown in Uttarakhand, only after his death. In fact, I had initiated this publication effort in the last days of his life, but due to his sudden death, the publication was delayed, and we could publish it only later in Hindi.

In all the movements in which he participated, Prasun made his special presence felt with the intensity of his feelings, his deep sincerity and his willingness to take up difficult risky work.

In his last years when more trees were being felled to make room for high-tension wires through the Garhwal region, Prasun and his friends revived the spirit of the Chipko movement and again hugged trees to save them. Finally, a committee of the Supreme Court visited the area and thousands of trees could be saved once again.

Prasun's activism thus continued till his very last days.

In fact, his entire life can be seen as a series of sincere and fearless experiments with various social and environmental movements. His integrity and honesty were always one hundred per cent. He gave his all to movements and for some time he tried to forget everything else.

Prasun was very modest about all his many-sided contributions. He was very enthusiastic when speaking about his work but carried his huge achievements very lightly. His attitude seemed to be – I am doing nothing great, everyone should be doing this!

What attracted me most to him was that he never appeared to yearn for 'fame' and for public recognition/acclaim of what he had genuinely contributed. While some other 'activists' exaggerated their contribution and converted this into international awards, Prasun was content to carry on his ordinary life full of daily struggles.

He had the most meagre personal possessions. Occasionally when handing him a payment for an article we had jointly written, I would see his face light up and he would tell me very frankly that he'll now be able to buy a tin of oil or other such daily necessity for his family on his way back to his village. Once, I wondered why he was so unduly agitated at his shirt getting drenched, till I realised that this was his only shirt. As far as I could see, he could never afford to buy a daily newspaper. He read newspapers mostly at tea shops in the local market and used a transistor to catch up with the latest news.

For me and for many others who were fortunate to know him closely, Prasun will always remain a source of inspiration.

In the early hours of 15 July 2006 Kunwar Prasun passed away at his family home in the village of Rampur (Tehri Garhwal district) in Uttarakhand. He was 56. He left behind his wife Ranjana, his two sons Anurag and Anupam and daughter Jigyasa.

If we have to sum up Prasun's life in just one sentence, then we can say that *He* was a *Man of Truth*. He never hesitated to speak the truth and always acted in accordance with his understanding of truth.

What was perhaps most remarkable about Prasun was the way in which the three major concerns/movements of our times – peace, justice and protection of environment – were integrated in his personality in a very effortless way. He was a living example of how these most important concerns are well-integrated components of the quest for a better world.

I first met Prasun when, as a freelance journalist, I went to his village to cover the Chipko movement. Later, our families also became very close. We will always remember Prasunji as a great friend and as a constant source of inspiration.

I can't resist the temptation of concluding on a personal note. Even among the closest friends, generally it is seen that there are at least a few occasions of tension or conflict, no matter how minor the conflict. My relationship with Prasunji was for about three decades and during these 29 years I do not recall a single instance of any tension or conflict between us, or even the remotest ill feeling due to any reasons. I've tried hard but I can't recall even a minor conflict in 29 years. I mention this not only because this relationship was so precious to me, but also because such relationships are becoming so rare in the present-day world that we need to preserve these memories.

Ghashyam Sailani

Prasun was keenly interested in folk-arts and folk music and well-informed about it. He really admired songs written by Ghanshyam Sailani which were widely used in people's movements. When Sailani died, Prasun was extremely sad. What is common to both these inspiring personalities is that their creative/writing work was so well integrated with their social concerns.

Sadly, Uttarakhand has forgotten Sailani all too soon. If society is to retain its capacity to struggle for justice and truth, it cannot afford to forget Prasun and Sailani.

Pandurang Hegde

Life Devoted to Environment Protection

IN THE FORESTS of Western Ghats, in northern Karnataka, a schoolboy grazing his family's cattle and buffaloes was deeply distressed. Before him was a sight he just didn't want to see – some beautiful trees had been cut and some others had big cuts on them.

This cowherd used to come to the forest with his family's animals in early childhood. Then he was sent to school in a town. He had now returned after a gap of a few months during school holidays.

While he was away, the trees in the forest near his village Kalgundikoppa (in Sirsi taluka of Uttar Kannada district, Karnataka) had been sold in an auction and these were now being felled.

The schoolboy cowherd did not understand the growing greed of the timber industry and the pressures to increase the forest revenues. But what he did understand was that these forests which were being felled had a very close and protective relationship with his family, his village, animals. This is something that he had gradually assimilated almost every day of his childhood as he grew up in his 35- member joint family in Kalgundikoppa.

What he saw every day of his growing up was that people of his village, his own family members, and neighbours were able to meet so many of their needs from forests which existed around the village. Without these forests and their trees, life in their village would be very difficult.

In addition to assimilating this knowledge, the schoolboy cowherd also simply loved the trees and flowers and the birds and animals that inhabited the forests. He loved to watch the playful parrots and hornbills, but most of all he loved to watch the flowing water, and the melody of the flow of natural springs in the monsoon months. Wearing the improvised raincoat made from hand-woven blankets for cowherds, this cow-herd would be immersed for long in the beauty of the flowing waters.

He loved also to feast on the plentiful summer fruits in the forest, and to bring home some of the tastiest vegetables which grew wild in the forest, particularly in the rainy season.

So, when he saw the sight of slaughtered and wounded trees in the forest, the cowherd himself felt wounded. He fell as if someone had attacked and hurt him.

At that time, he only felt the deep anguish, a feeling he knew was shared by other villagers. He didn't know how to check the cause of this distress. At that time, he did not know if the felling of trees could be stopped.

He carried the scars of the wounded and weeping trees deep in his heart, but he did not know that one day he'll be able to do something to prevent this tragedy. He did not know that one day he himself will play a leading role in checking this slaughter of trees.

At that time, of course, this cowherd named Pandurang Hegde could not imagine that he'll one day be a leading activist and eventually the coordinator of the Appiko movement for the protection of millions of such trees in the State of Karnataka and in other parts of south India.

Born as the fifth child of Suman and Ganapathi Hegde in 1956, Pandurang Hegde grew up in a joint family in the village of Kalgundikoppa with several brothers, sisters and cousins. His father had been a leading freedom fighter of the area. Pandurang went to the village government school and also helped with farm work. He felt particularly keen to help with farm animals and taking cows and buffaloes out to graze.

Recalling those days, Pandurang says, 'To be a cowherd is an invaluable educational experience. This work involves forming an intimate relationship with the animals you graze, for without such a relationship this work cannot be done peacefully'.

'But once this relationship is formed, the cowherd can work in a restful way for the animals know not to wander too far and when to return. They respond to the call of the cowherd, and the cowherd knows what kind of sounds to make for what purpose. All this involves building a relationship of understanding and trust with the animals, which is very valuable education difficult to get in other ways.'

Musing about his childhood relationships, Pandurang adds,

'People generally regard cows as sensitive animals but do not have the same opinion of buffaloes. But my daily interaction revealed how buffaloes too are very sensitive animals. In the extended joint families, various members and particularly children would form fond ties with particular animals and these animals too would respond to them with the same affection'.

This village life was very isolated then due to lack of communication. Pandurang would go to the nearest town of Sirsi (16 kms away) generally only once in a year, walking considerable parts of the distance.

Pandurang's idyllic life in the village was disrupted when his father died tragically at the age of only 52 in 1962. Pandurang was only 6 years old then. After some time, his elder brother, a government employee, took him to Nagpur for further studies. Pandurang pursued his further education in Nagpur and graduation in commerce form Gogte Institute in Belgavi in north Karnataka, but he regularly came back to his village for all holidays.

The early days of cowherding were later followed by bigger responsibilities like mango harvesting. Pandurang says he has harvested as many as 40,000 mangoes in one season, but none of these was sold, all were shared with friends, relatives, visitors and other villagers. This learning too has stayed with him, even today he doesn't sell the chikus and mangoes of his farm, and these are only for sharing.

Despite being educated in several towns and cities, Pandurang's links with his village remained very strong. Problems of villagers and their sufferings were close to his heart. He quickly grew tired of his accounting education at the college level and he longed for a different education in which he could link up again with his rural roots. So, when opportunities emerged for a Master's Course in the Delhi School for Social Work, University of Delhi, he responded with alacrity and somewhat to his own surprise, his application for the course was duly accepted.

Encouraged by his mentor Prof Shankar Pathak at the School of Social Work and his wife Prof Zakia Pathak (who taught English literature at Miranda House), young Pandurang took his educational opportunities very seriously and used his holidays to travel all over the country (with hardly any money in his pockets).

He lived with diverse kinds of people and learnt from them. When he did not get any other place to sleep, he slept at the railway station with porters. Recalling those days, Pandurang says, 'Wandering without any specific aim and meeting various people in a free and uninhibited way has been a very useful experience for me'.

In addition, there were more specific placements with social work organisations – in pshychiatric social work at the All India Institute of Medical Sciences in Delhi, and Community Development work in the slum areas as a part of his formal education.

He benefited much also from his love for literature, thoughtfully devouring the novels and stories of Kannada writers like Shivram Karanth, Kuvempu and others who documented so vividly the reality of the Malenadu (hilly) region from where he came. In addition, he also became fond of Hindi and English literature. At a very young age he could read, write and speak four languages – Kannada, Hindi, English and Marathi.

The hilly region links made him choose the Himalayan Uttarakhand region for the field-work component of his post graduate education. This was the region where the Chipko movement had taken roots. Pandurang went to senior Chipko activist Sundarlal Bahuguna's ashram in Silyara village of Tehri Garhwal. Here he formed a lifelong bond with Sundarlal, his wife Vimla and their senior associate Dhum Singh Negi. In addition, he made numerous other friends from the Chipko movement, particularly those based in Tehri Garhwal district such as Chipko poet and folk-singer Ghanshyam Sailani.

Destiny had brought Pandurang from Karnataka to these Himalayan villages and Chipko activists at a very young age – it is these linkages that were to soon shape and guide his eventful life.

Those were very inspiring times in Tehri Garhwal as in Henwalghati and other places, women and other villagers had protected forests by hugging them. These efforts had succeeded in protecting several forests like Advani and Salet, and there was much expectancy that the government will agree to the Chipko movement's demand for a much wider ban on green trees felling in Uttarakhand (which was at that time a part of Uttar Pradesh). (This demand was subsequently actually met and in addition, Uttarakhand also became a separate State).

Pandurang carefully listened to the many inspiring stories related by activists and villagers and took part in several padyatras (or foot marches) organised to spread the message of Chipko movement in more and more villages.

He later also participated in some stretches of the long foot march taken up by Sundarlal Bahuguna from Kashmir to Kohima. These foot marches were taken up without any cash support and were entirely dependent on the support people of any village agreed to extend to marchers. All this was invaluable training for Pandurang's future work in his own hilly and forested region in Karnataka. By the time he returned to the Delhi School of Social Work, Pandurang had been accepted as a member of the Chipko family in Tehri Garhwal district.

When Sundarlal Bahuguna came to Delhi and Pandurang went to meet him, the senior activist took him to meet several prominent persons including media persons and Gandhian leaders. On some occasions, Pandurang as a young disciple also accompanied Sundarlal Bahuguna when he went to meet the Prime Minister Indira Gandhi with the demands and messages of the movement (They met Indira Gandhi when she was Prime Minister but also when she was out of power).

Pandurang not only got his Master's degree from the prestigious Delhi School of Social Work but also emerged as a gold-medalist. He had several offers from the corporate sector but decided to ignore this to continue his aspirations of grassroots work. A suitable opportunity emerged in an Oxfam supported programme for rural landless workers, and Pandurang found himself in Damoh district of Bundelkhand region (Madhya Pradesh, central India), working with his team-leader Andrew Clark to help set up the Damoh Gramin Shramik Kendra, an organisation of rural workers which still continues under a different name.

Andrew Clark was an inspiring leader with whom Pandurang could share many concerns and values. Andrew named his house in UK Chipko. The programme progressed well; but a time had to come soon to hand over responsibilities to the emerging local leadership who had a better grasp of local realities. So, after about 3 years in Damoh, Pandurang headed back to his native 'malenadu' – Sirsi taluka of Karnataka.

Pandurang roamed around with many Chipko ideas in mind but with little hope of translating these into real actions.

A padyatra along Kali river was a very useful experience, but this did not yet involve others as co-marchers. To some local people, he was beginning to look like a strange person who had fled from good opportunities after completing his education and was now just messing around here and there, doing nothing in particular.

This perception of Pandurang was strengthened among several people when a letter came – of all the places – from the Prime Minister's office (Indira Gandhi had by then been re-elected as Prime Minister) informing Pandurang that he had been selected for a post in the personal staff of the PM and should soon report for duty. Many people including his genuine well-wishers thought Pandurang must be half-mad to reject such a grand offer at such a young age. But Pandurang was firm that his path lay elsewhere, and he was not about to accept the generous offer from the Prime Minister's office.

Recalling the incident, Pandurang muses that Mrs Gandhi must've remembered him from the meetings, where he had accompanied Sundarlal Bahuguna to discuss environmental issues. This can be the only explanation why such an unexpected offer came at a very young age. Probably the Prime Minister intended to give some responsibility for environment protection to him.

Anyway, there was no time for such guesses as some events suddenly started moving at a brisk pace. Reports of excessive tree-felling were creating increasing anger and frustration among people as with the destruction of forests life would become extremely difficult for villagers.

As youth clubs had been active in several villages of Uttar Kannada district in fighting several social problems, some of them also took up this deforestation challenge. A particularly active youth group in Balegadde village had been active in drawing attention to excess felling in forests and writing to the authorities to protest against this. As Pandurang had been going around telling stories of the Chipko movement, an idea emerged at youth club meetings that Sundarlal Bahuguna should be called to the village and Pandurang Hegde should help in arranging this visit.

Now Pandurang suddenly had a feeling that he was in the right place at the right time. As Sundarlalji was visiting Karnataka around this time, it was not difficult to call him to Belagadde. The events were now moving fast towards the launching of Karnataka's, and south India's first movement for saving forests. As Sundarlal Bahuguna said, 'Like a bird the Chipko movement flew from the Himalayas to reach the hills of Uttar Kannada'.

Today, Pandurang Hegde is an internationally known and respected figure.

The remaining story is told in this book in the Chapter on Appiko movement. The movement for saving forests went from one achievement to another, overcoming many problems, and Pandurang made a very important contribution to this entire effort.

Radheylal Herlal Richharia

Sacrificing All to Protect Farmers

PROF PS HUDSON, the venerable botany scientist at Cambridge, was just settling down to some long painstaking work at his lab when the sudden appearance of an earnest young man startled him.

The Professor scratched his head. Yes, he remembered having seen the youth somewhere, but exactly where he could not recall.

'Sir', the young man sought eagerly to refresh his memory, 'I hope you remember that when you came to Nagpur some time ago, you had praised my work and invited me to come to Cambridge for my doctorate'.

Oh yes, the moment came back to the Professor in a flash. He had been visiting some labs in India when he chanced upon this brilliant student Radheylal Herlal Richharia who was doing his MSc at Nagpur. The student had been singled out to Prof Hudson as his original scientific work had already won appreciation at prestigious fora of scientists.

Interested, Prof Hudson had taken a close look at some of this student's work. Then, in a gesture of appreciation of the work he had said, 'You must come to me at Cambridge for your doctorate'.

But after this there had been absolutely no correspondence, no follow-up, no indication at all that this student was actually planning to come to him for his doctorate.

'I have come without any papers and without any formalities, with just your words to support me. I did not have any money to come here, nor do I have money to stay here. I could come only because I passed an exam for a scholarship which is meant to send students for ICS (Indian Civil Service) exams', the student explained.

So here was a brilliant student very keen to do scientific work but with a scholarship for an entirely different purpose. The problems in admitting him for doctorate were immense, but Prof Hudson also hated to lose talent to the ICS.

So, he told the student that he could start working at the lab, pending a final decision on admitting him for doctorate at Cambridge.

This was all the encouragement the youth needed. He said goodbye to his colleagues who had come with him to appear in the ICS exam, and settled down to rigorous work at the lab, working all the day into the night hours, just as he had done at Nagpur.

Unfortunately, the new cold environs did not suit him. He soon had an attack of pneumonia. But this did not did not prevent him for coming to the lab and pursuing his work with almost the same rigour as before. His zeal as well as the excellence of his work as revealed in the preliminary results, did not escape his teachers.

Despite the absence of many formalities, he got his admission for the doctorate course at Cambridge.

In less than two years he had obtained his doctorate.

Unfortunately, however, he did not have the means to return home. The Indian High Commissioner graciously stepped in to help. And at the age of 25, Radheylal Richharia was back among family and friends in India with his richly -deserved doctorate from Cambridge.

His Professor from Nagpur, who had called him mad for venturing to even think of getting his doctorate from Cambridge, now had to reconsider his opinion.

Several promising avenues were now open for the young scientist. But he remembered every word of the parting advice he had got from his senior teacher at Cambridge. Before leaving, he had gone to say goodbye to Prof Engledow who patted him affectionately and said, 'So many scholars who return from Cambridge to India go on to hold many prestigious posts and to publish many learned papers. But I do not know how many of them actually strive to help the farmers of India. The real test is whether your work will be of real help to the poor farmers of your country'.

These parting words from the seat of learning which he valued so much made an indelible impression on the young scientist's mind. In the midst of all his future struggles and achievements (both of which he was to experience in plenty) he would never forget these words – your work is valuable only if it really helps the poor farmers.

Radheylal was born in 1909 in Nandanvara village of Seoni Malwa tehsil in Hoshangabad district of (what is now) Madhya Pradesh. His father Herlal Richharia was the headmaster of a school and also the postmaster. Always keen to do some new work, he experimented with growing several kinds of vegetables in his garden.

But he did not grow vegetables just like that. He had careful and painstaking methods of growing the various vegetables. In some cases, he would not use any direct irrigation at all, preferring instead to use the percolation method, saying that this way the land would make more efficient use of water. Then in the case of some vegetables which grow in creepers, he would insist that the creeper be given the support of a neem tree. The belief probably was that the vegetables would absorb some of the undoubted medicinal properties of the neem plant.

All this while his curious son looked on eagerly at his father's work, fascinated by all the interesting information that was revealed when he asked questions. The child learnt to admire and respect the wealth of wisdom that was available within the villages.

He was equally fascinated by the visitors from forests, or adivasis, with whom his father had great rapport. Specially fascinating was the 'Rani Sahiba', who seemed to have been a tribal head woman of sorts, with whom his father appeared to have an arrangement for care of animals. It appears that villagers then had an arrangement with some forest people to take their domestic animals to the forests during the dry season. The tribals also collected chiraunji seeds and not only brought this to his father, but also prepared beautiful storage for keeping them unspoilt. What the adivasi visitors said revealed their rich wisdom of trees and plants.

Radheylal had learnt of this wisdom at an early age, and also the memory of the friendly relationship of mutual help between his father and the friends from forests lingered on in his memory.

His father watched the growing interest of his son in plants and visitors and encouraged him, allowing him to sit and listen to the conversation with the tribal people, or taking him to the market to meet them. Little did he realise that his little boy will grow up to become one of the biggest defenders of tribal heritage, and that some of his valuable work and publications would be dedicated to the tribal farmers of India.

Before he left his village for higher education, Radheylal had imbibed enough knowledge about the wisdom of farmers in general and tribals in particular, a knowledge that would enable him to retain his humility and attitude of respect for them even when he approached them after earning his doctorate from Cambridge.

Radheylal was such a brilliant student that his officially-recorded date of birth had to be lowered by two years (from 1911 to 1909) to enable him to sit for school exams for which he was obviously well prepared, but for which he was still well below the required age. In fact, at one exam he had to be helped to climb the seat which had been designed for older students.

Radheylal soon left his village to pursue his further education at Balaghat, then college-level education at Varanasi and Nagpur. From Nagpur he went straight to Cambridge and it so happened that on returning from Cambridge, the first job that he got was also at Nagpur-as oil seeds specialist for Central Province (1931-42). From here he went to the Agricultural Research Institute and Agriculture College at Sabour (District Bhagalpur, Bihar) where he held various positions and developed his keen interest for rice which was to become a lifelong passion. He also showed his flair for field-level work compared to desk work by motivating students to do the actual work in fields and dairies, thereby earning themselves a rich breakfast of halwa (a sweet dish) and milk.

On a more serious note, he started pioneering work on preparing quality fibre from the waste straw of linseed plant, a work which drew the attention of several eminent persons. He remained in Bihar for 17 long years, from 1942 to 1959.

Then in 1959 he was selected as Director of the Central Rice Research Institute (CRRI) at Cuttack, Orissa. No sooner had he reached there to assume charge, however, a rival who wanted the post for himself manipulated a strike by employees so that he would not be able to settle down in the new work. This hurdle was soon overcome, however, and RH Richharia was able to establish the CRRI as one of the foremost centres of rich research in the world, the value of work done there was widely talked about. He made preparation to launch a very well-planned effort to improve rice cultivation based on indigenous seeds.

However, vested interests conspired against him to disrupt his work and he left the CRRI in 1967 and despite his seniority, was unjustly deprived of the opportunity to become the Director General of the Indian Council of Agricultural Research (ICAR). Instead, a programme based on exotic seeds and imported technology was launched which disrupted eco-friendly traditional farming practices very badly. The imported Green Revolution nudged him out.

In 1971, he was recalled from retirement to head the Madhya Pradesh Rice Research Institute at Raipur and also to function as Agriculture adviser to the Government of Madhya Pradesh. His work here from 1971-1976 was again pioneering work of great value with special reference to preserving and protecting the agricultural heritage of tribal farmers.

He collected a vast range of indigenous rice varieties and was again on the verge of a very promising programme of improving rice cultivation based on these indigenous varieties. Again, however, vested interests manipulated things to deprive him of opportunities to continue work of great value which had reached a critical stage.

From 1978 to 1996, RH Richharia, now deprived of all facilities for scientific work, continued his work against all odds at his residence and at a farm located a few miles away from his residence in Bhopal.

In 1971, Richharia lost his wife Jaiamba Richharia, who had been his greatest source of strength in the middle of several difficult struggles. Around 1982 he lost a son. And in 1983, he was requested by Prime Minister Indira Gandhi to prepare an action plan for improving rice production. This he did with great dedication but following Mrs Gandhi's assassination, this plan was ignored. In 1984 he became one of the victims of the Bhopal gas tragedy, the man who had spoken all his life against the hazards of chemical pesticides and distorted development trends himself became the victim of one of the world's greatest chemical disasters. In December 1984 he had a heart attack, the after-effects of which left him very weak.

But he withstood all these tragedies and shocks to continue his work.

His daughter Utpala combined a busy professional life with affectionate care of her ailing father. Other members of his closely-knit family, particularly elder daughter Anuradha, also helped to keep up his spirits after all the harm he had suffered from the hands of manipulators. Several of his old students scattered all over the research and teaching establishments also extended moral support from time to time. They kept him in touch with the recent developments in research and confided to him how the value of what he had said and done much earlier was being realised now.

Slowly, information on his work and realisation of its importance became clear to several members of the press, voluntary agencies and other people's organisations. Reports of his work started appearing in several newspapers, magazines and journals. Voluntary agencies requested him to visit their areas and tell people about his work.

He was invited to important gatherings of pro-people Science in Malaysia, Indonesia, Nepal and elsewhere. He told people how he had always believed in working closely with farmers and in the process, he could learn so much from them. He always kept in mind their welfare and never wavered from this, no matter how hostile the conditions.

Despite numerous adversities, he was able to write and publish a very important book on clonal propagation technology. A new edition of his old book, *Rices of India*, co-authored with S Govindswamy was published.

Radheylal Herlal Richharia died in 1996 at the age of 87.

In his last days, Richharia was working on an encyclopaedia of rice germplasm of Madhya Pradesh (then including Chattisgarh) in which 17,000 cultivars of rice were being listed. This is the result of several years of the collection and identification of rice cultivars, a work which he continued even after his retirement from official work.

He was also completing the botanical survey work with special reference to rice and rice research conducted by him during the seventies in MPRRI.

He was also continuing his tireless work in preparing a Botanical Dictionary of useful plants in India. This work was started in 1946-47 and over 8,000 entries had been made (with available chromosome family name, botanical name, popular English and Hindi names, uses and other information).

In addition, he continued to carry out botanical research on his farm and had been working or evolving interesting concepts such as those of rice garden and rural gene banks.

All this work he had been doing with the minimum of help and support from outside. Some help, however, was extended by agricultural universities in Madhya Pradesh and by some voluntary agencies/institutions by agreeing to publish a part of his voluminous work.

In the course of his careful, painstaking and brilliant research work extending over six decades, RH Richharia's work was distinguished by some important features which are:

1. A respect for wisdom and knowledge of ordinary farmers, especially tribal farmers.
2. A deep concern for the small and poor farmers, an understanding of their resource constraints and efforts based on this understanding to make available low cost, local resource-based technology. His technology involves no chemical pesticides.
3. A respect for the indigenous genetic base of rice, and a great concern to preserve the indigenous germplasm.

Efforts to increase rice productivity should be based mainly on indigenous varieties of rice, he advocated. Towards this end he made a great discovery of the hidden potential of indigenous rice varieties, their high productivity and superb food qualities. In Madhya Pradesh(then including Chattisgarh) he collected (or guided the collection) of over 17,000 cultivars of rice.*

https://satavic.org/dr-richharias-story-crushed-but-not-defeated/
https://www.thehindubusinessline.com/todays-paper/tp-others/article29028934.ece
https://www.thehindu.com/sci-tech/agriculture/rice-formula-on-the-back-burner/article5225869.ece

Mangal Singh
Devoted Farmer-Scientist

MANGAL SINGH IS a farmer of Bundelkhand region who was known at a very young age for his enterprising and innovative farming methods.

Mangal used to see regularly that small farmers with low resources have a lot a difficulty in buying diesel for lifting water from streams and rivulets. He started thinking a lot about the possibilities of lifting water from small rivers, nullahs and canals without having to depend on diesel or electricity. After a lot of thinking and experimentation, he came up with a device which could lift water using the energy of flowing streams without depending on diesel or electricity.

This invention was named Mangal Turbine. It was first demonstrated in 1987. Mangal Singh was 40 at that time. Later it was patented as 'Mangal Water Wheel Turbine Machine'

(Patent No. 177190 dated 13-11-1997) as per Government of India gazette Notification dated 30th November 1998.

This technology is described by Mangal Singh in the following words:

> The water wheel turbine machine consists of a water wheel which is firmly mounted on a steel shaft and supports on two bearing blocks fixed on foundation supports. The shaft is coupled with a suitable gearbox through universal couplings for stepping up speed of rotation. Output shaft of the gear box is coupled on one end with a centrifugal pump for lifting water and the other end is mounted with a suitable pulley for deriving power for operating any machine. Design of the water wheel turbine is simple. It is available in different size to meet the varying requirements. Operation of water Wheel Turbine Pump-cum-P.T.O. Machine is very easy as anyone can operate the machine by opening the wooden or steel gate valve, the machine is stopped by stopping the flow of water through the gate

Thus, apart from lifting water the Mangal turbine can also be used for several additional tasks. In the words of Mangal Singh,

> This is used for pumping water from the rivulets and water streams on which it is installed. The machine can be used for several rural works such as operating attachakki, sugarcane, crushing, threshing and winnowing, oil expelling, chaff cutting, etc. The machine provides a clean alternative (non-conventional) source of energy in remote rural areas for increasing agricultural productivity, income and employment.

By linking it to a generator, this machine can also provide electricity.

The Maithani report has explained the idea and functioning of Mangal Turbine:

> Mangal Singh of village BhailoniLodh, Block Bar, district Lalitpur UP invented a fuel less water lifting device in 1987. This device is known as 'Mangal Turbine'. There are several variants of the Turbine but the standard device contains a water wheel of 2 meter diameter with 12 blades radially fixed to the rim. The shaft is coupled with a suitable gear box for stepping up of rotation to 1500-1800 rpm. The output shaft of the gear box is coupled on one end with a centrifugal pump for lifting water and the other end is mounted with a suitable pulley to operate any other machine like crusher, grinder etc. By using the energy of flowing water in a stream, Mangal Turbine enables lifting of water for irrigation/drinking purposes and also produces mechanical power that can be used for various other purposes.

Several highly placed officials who examined Mangal Turbine carefully confirmed its great value and utility.

At least two former Secretaries of the Rural Development Ministry (BK Sinha and SR Sankaran) were known to be admirers of Mangal Singh's efforts. Sinha has spoken widely about the great usefulness of MT. Others who have admired and appreciated the MT include the Alternate Hydro Energy Centre IIT Roorkee, The Energy Research Institute (TERI) New Delhi, Punjab Singh, former Director IARI Delhi, Senior bureaucrats like Kalika Prasad, the then Commissioner Jhansi, Sarla Gopalan, former Advisor to the Planning Commission New Delhi and RS Tolia former Director State Institute of Rural Development-Delhi and MC CAPART, Central Zone Lucknow.

BK Saha, former Chief Secretary, Govt of Madhya Pradesh, has written about this device: I made a detailed analysis of the economic viability of the 'Wheel' and its comparative advantage vis-a-vis alternative methods of pumping water from streams and small rivers for irrigation. The system is extremely cost-effective even after taking into consideration the cost of the Stop Dam. Where the Stop Dam is already available, the system is even more cost-effective. Installation of this device is strongly recommended wherever there is flowing water in small streams by constructing a stop dam and installing one or two water wheels as designed and developed by Shri Mangal Singh. It saves on energy like electricity or diesel and is ecologically completely benign.

Similarly, TP Ojha, former Deputy Director General (Engineering) of the Indian Council of Agricultural Research wrote that: Mangal Singh's device offers great promise and possibility of lifting river water for irrigation, fisheries, forestry and drinking purposes. The water head created by putting a check dam across the river or perennial water course generates enough force to rotate the water wheels to operate one or two centrifugal pumps in series.... The designer of the system Sh. Mangal Singh deserves appreciation and support to install few more units in different parts of the country for demonstration purposes.

A report titled,'Problems and Potential of Bundelkhand with Special Reference to Water Resource Base' was prepared in 1998 by the Centre for Rural Development and Technology (CRDT) IIT Delhi and Vigyan Shikshan Kendra (VSK). This report examined MT carefully and recommended it for its great utility. This report said, 'Most significant aspect is that the entire system designed by Mangal Singh is easily fabricated in the village itself, using available material and local workmanship. Besides, it requires minimal maintenance compared to other types, expertise for maintenance is available in the village itself'.

'Using his engineering skills (through he had no formal

engineering training), Mangal Singh coupled a sugarcane crusher to the main shaft using a belt drive. Simultaneously, both water pumping and crushing could also be done. Similarly, the energy generated could be used for running a grain thresher, grinder etc. He uses this energy for operating the machines tool of a local workshop. Effectively water wheel becomes a source of rotational energy which can be used for any purpose.'

'Thus, Mangal turbine would prove a boon for fulfilling the energy need of irrigation, agro-processing etc. in the rural sector wherever low water head exists in the rivers/Nallah.... Preliminary study of this system conducted by IIT personnel has indicated that its efficiency can be further improved by using some modern scientific-technical inputs.'

'This turbine is a fine example of common people's inventiveness, and should be encouraged by all means for people's benefit. It is unfortunate that in the pervasive atmosphere of "foreignomania", this device has not got the recognition it deserves.'

The Maithani Report says about Mangal Turbine:

> It is undoubtedly unparalleled in its simplicity and utility. Its cost benefit cannot be restricted to the extent of area irrigated and increase in production and income on account of that. Its benefits are multiple and multidemensional. Bundelkhand is a drought prone area and its main problem is lack of irrigation. Unfortunately our policy makers and planners prefer big and extravagant projects which allow pilferage and splurge. Mangal Turbine offers the low cost, environment friendly and sustainable solution to the irrigation problem of Bundelkhand.

SOCIAL CHANGE AND THE SELF

If you want to change the world, start with yourself.

—Mahatma Gandhi

When you are right you have no need to be angry. When you are wrong, you have no right to be angry.

—Mahatma Gandhi

Changing Ourselves
Also Changing the World

THERE ARE MILLIONS of life-forms on earth – mammals, birds, insects, reptiles, fish and others. However, human beings alone have certain distinct abilities which no other form of life has. Human beings alone have the capacity to work in a planned and organised way for the protection of various forms of life (existing presently and their future generations) as well as the eco-systems and habitats which support these life-forms.

Human beings, of course, have a duty and capability to help and protect other human beings (members of their own species) but in addition they alone have the ability to work in an organised way to reduce the distress and threats faced by other, very diverse life forms. This unique ability of human beings defines their main role. The true role of human beings on earth is to help each other and protect other forms of life as well as their habitats, now and in future generations.

A person's life lived in accordance with this understanding of human life may not be an easy life but it is likely to be a very happy and peaceful life as any sincere involvement with reducing the distress of others brings a deeper happiness and peace which cannot be experienced in any other way.

It would have been very nice, our world would have been a very happy place if all human beings had inherent qualities to live in accordance with this role of human beings. In reality, however, human beings are (at least) as likely to live in gross violation of this desired role as they are likely to live in accordance with it. In any given time in a society, the extent to which people live according to this role will depend on the kind of influences to which they are exposed in any society, particularly in family and school or by media.

It is possible that at any given time in a society the dominant influences are such that most people are motivated mainly to get as rich as possible, to grab as many resources as possible regardless of the adverse impact of this on other human beings and other forms of life. In such a society, most human beings are likely to be far, far away from the true role of human beings, leading to very high levels of distress for most people as well as for other forms of life, for this generation as well as for future generations. In such a situation instead of fulfilling its role as the protector of all forms of life on earth, humanity unleashes a lot of distress on its own members while at the same time destroying on a massive scale other forms of life, the habitats and eco-systems which support them.

This is the true situation today, and the destruction is on such a massive scale that the very survival of many forms of life is badly threatened. Weapons of mass destruction have accumulated to a point where these have the destructive capacity to wipe out human beings and most other forms of life. Climate change threatens a rise in sea level, the submergence of many coastal areas, the recession and disappearance of glaciers, an aggravation of several kinds of disasters and appearance of new diseases against which immunity may not be available. All these forms of destruction are rooted in man-made causes. The need for human beings to realise the true goal of human life and remain close to it was never greater than in present times. Any delay at this juncture in coming closer to true role will be simply too expensive in terms of the destruction unleashed.

So, as a first step it is very important to set out the conditions in which human beings are most likely to remain close to their true aims, their main role on earth.

Firstly, it should be possible for all human beings to meet their basic needs so that they can devote their creative energies to realise their true role of helping other human beings and other forms of life in this and future generations. This requires the setting up of just socio-economic systems among human beings.

Secondly, the significance of remaining committed to the true goal of human life should be emphasised in family, in school and college, in media and all other important institutions and fora of human life. As a person grows, the external environment on the whole must be favourable to realising the importance of the true goal of human life.

Thirdly, adequate opportunities for such education and training should be available which increase the capacity of human beings to remain committed to the true role, while avoiding the at-times attractive looking path of getting entirely submerged in the pursuit of sensual pleasures and material gains.

Societies and times which take care to provide these three conditions will be rewarded by a much higher percentage of people remaining committed to the true goal of human beings.

This entire effort of course is a voluntary one, based on motivating and inspiring people and not on forcing or ordering them.

Thus, a most significant, we may see critical task of our times is to create conditions in which more and more people feel inspired to lead their lives in accordance with the role of helping others and other forms of life.

Practical Life and Basic Role

Now let's look at this entire question of remaining committed to the true role of humankind from a more practical viewpoint of an individual. This person agrees with the principle that the main aim of his/her life should be to strive to protect and reduce the distress of other forms of life as well as help fellow human beings. Now how to realise this in actual practice.

Let's assume that the name of this person is Roshan and he lives in Delhi. One challenge for Roshan is to find a source of livelihood which is in keeping with this aim. This may be quite difficult, as most of the economic activity in the present-day world is not attuned to this need. If he is lucky, Roshan may yet be able to find a satisfactory source of livelihood but probably at a lower salary than what he could've obtained elsewhere. Or else he may be able to innovate some sort of self-employment which is in keeping with the main aim. Even if such options are not available and Roshan is forced to take up employment which is not in accordance with the main aim of his life, he'll accept this as a temporary compromise till he can find more satisfactory arrangements.

It would've been great if he had a full-time job in accordance with the central aim of his life, but even without this advantage Roshan is determined not to lose sight of the central aim of his life. In his inter-actions in his family, his neighbourhood, his working place he always keeps this in mind, striving to be as helpful as possible particularly to those in distress and need, whether it is a victim of accident in a nearby slum or a stray cow abandoned near his home. In his spare time, he volunteers to work for a campaign on the rights of landless peasants, and for another campaign on cruelty against animals.

As a citizen, Roshan carefully keeps in touch with current events and raises his voice against injustice wherever it is possible and practical for him to do so. He is careful to use his vote in various elections in such a way as to help the cause of justice. He uses whatever opportunity is available to him to inspire more people to live in accordance with the true role of human beings on earth.

In making various choices in his life Roshan keeps in mind the welfare of other people and other forms of life. For example, consider such a simple decision as - what shall I eat and drink? Like most people Roshan of course likes to eat what he relishes, but he exercises his choice of food within some self-imposed conditions which are in conformity with the central aim of his life. For example, his food habits should not involve injury to other forms of life, should not involve a wasteful use of food and land, should not lead to bad health or intoxication. In a positive way, Roshan will be willing to make an extra effort to obtain food which, while contributing to his health and nutrition, is also linked to helping poor peasants, protection of farmland and rural environment. All these considerations lead him to select a basket of food which avoids alcohol, tobacco and meat, reduces foods rich in fat and sugar, while encouraging the use of organic grains, legumes, vegetables and fruits.

Similarly, when he makes other choices (such as where I will live and what clothes I'll wear), Roshan of course looks at his own comfort and security but also considers whether his choice can help those artisans or workers who need to be supported, and whether he can save resources for the higher, central purpose of his life.

Roshan may be working harder then some of his friends and having lesser material comforts than them, but he is at peace with himself and frequently experiences the unique, deeper happiness that can only come from helping to reduce the distress of others. His life contributes every day, no matter in how small a way, to the welfare of other people and other forms of life. His life is based on very clear understanding of how human life should be lived and so, despite numerous difficulties and dilemmas he daily faces, he is at peace with himself and with the rest of the world.

Grassroots Efforts

A person with such an understanding and conviction should, of course, strive to spread this to other persons. This effort can start closer at home with one's own family. The effort should be to create conducive conditions in which all family members acting voluntarily are drawn towards the true aim of human life. The family decides in a democratic way how at a practical level, they can live a life in accordance with this aim of humanity so as to make the most creative and productive use of their abilities while also meeting their needs. The quest for realising the goals of human life becomes more satisfying, inter-active and exciting in the company of one's near and dear ones.

The love and kindness which various family members have for each other are significantly extended, as the entire family reaches an understanding to have a similar relationship with other human beings as well as other forms of life. Their efforts on a continuing basis to reduce the distress of other forms of life enriches then own life and strengthens their own bonds. These efforts gain from the creative contributions of several persons.

Of course, such an ideal state is not always possible. Even small, slow steps together on the path towards the main goal of human life are most welcome. In situations where even this is not possible, efforts at the individual level can still continue.

The person who is convinced of the main role of humanity, of course, tries to take this message to other people, wherever this is practical and possible. As the number of people convinced about the true aim of humanity increases, a number of people can get together to form a group. The members of a group help each other to live a life in accordance with the main aim. They can also work together to take up larger responsibilities. They can also start various ethical production and trade activities so that their members get more opportunities to have sources of livelihood which are in accordance with the main aim of humanity. As the number of such groups increases, they can co-operate with each other to take up yet wider responsibilities.

An essential feature of these groups is that they have a very broad vision of the purpose of human life and their various activities, including single-issue campaigns, exist only as part of this overall understanding. This is how these groups are distinct from those groups which are largely limited by the pursuit of single issues or campaigns of a narrow scope.

Political Power

As the number of such people, families and groups continues to grow, a question arises whether they should also aspire to some form of political power, extending all the way to forming their own government, by contesting elections at various levels (in villages and towns, at the provincial and national level). Wherever such democratic opportunities exist such an effort can certainly be made as long as it is possible to keep it a highly principled, honest and transparent effort which is entirely free from the corruption and opportunism which trouble most political parties these days.

The distinct feature of this political party should be that it should devote itself without any selfish motive to formulating a programme based on reducing as much as possible the distress of all human beings (without any discrimination whatsoever) and other forms of life.

This political party should propagate this programme among people and ask for their votes on the basis of this programme. When the people are ready for this programme, they will vote this political party to power. Till that time, this political party should be content to play the role of an opposition party and also devote its energy to ensuring that the election process becomes as fair and free as possible.

The tendency to dilute or sacrifice principles, just to gain votes, should be completely avoided. This is not to say that the party's programme should be rigid and not open to necessary change, but only to emphasise that change should be only on the basis of a better understanding of the basic principle of reducing distress of human beings and other forms of life.

Secondly, all transactions of this political party should be completely transparent and honest. Accounts of all income and expenditure should be carefully maintained, and these should be open to inspection by citizens. For contesting any election, financial resources should be raised as for as possible by citizens/supporters within the constituency. All expenses should be kept as low as possible. Elected representatives from this political party should be encouraged to share their salary, housing and other benefits with the political party, and set an example of frugal, transparent life-styles.

If this political party is able to form a government in a country, it should work with the greatest dedication to provide a working model of a government which is committed to reducing the distress of human beings and other forms of life, now and in the generations to come.

World Government

This vision of change should extend beyond a few isolated countries to the concept of a world government. Those people who are committed to reducing the distress of all human beings (without any discrimination) and other forms of life are the most qualified to initiate and spread the vision of a world government.

An immediate need for world government has arisen due to the aggravation of some trends (such as climate change and proliferation of weapons of mass destruction) with the potential of causing massive worldwide destruction.These can be called survival issues.

As, at present, the concept of nations with their narrow boundaries is well-entrenched, a world government can be formed without disturbing the framework of these nation-states for the purpose of only tackling global emergencies such as climate change and weapons of mass destruction. For this purpose, elections can be held for a world parliament supported by best available experts to advice on survival issues.

However, at a later stage the jurisdiction and functions of world government can be extended. This will need some years of preparation in the course of which more and more people should be encouraged and motivated to think in terms of the welfare of all life forms of the entire world instead of thinking only for narrow nation-level interests. Persons who have already accepted reducing the distress of all human beings and all forms of life as the central role of humanity will no doubt be expected to play an important role in bringing such an important change.

Religious or Secular Approach

What is the role of religion in this entire effort?

The most significant aspect of this change is that more and more people should accept reduction of distress of human beings and other forms of life as the main aim of human beings which defines their role on earth. It is not easy for people to lead a life in accordance with this role. It is much easier to be tempted by a life dominated by the pursuit of comforts and sensual pleasures.

So, a person may need to draw inspiration or strength from some source. Religion can be an important source for many people for such inspiration. Of course such an inspiring concept of religion should be entirely free from bigotry and hostility towards such faiths.

Apart from religion, however, there can be other sources of inspiration such as a great individual or sect in human history. A person can also obtain inspiration from his/her parents or a friend. Hence this path is as open to deeply religious persons as it is to secular persons and atheists. A person who follows this path is free to derive his/her inspiration from either religion, or a historical figure or his/her own friends or the person may have enough strength within so as not to require any outside source of inspiration at all. If religion is to be the source of inspiration, then this can be any religion which doesn't base itself on hostility towards other faiths.

Anyone, Anyplace

This, then, is a vision of change which has the advantage of being adopted by anyone, anytime, anyplace and yet also moves in the direction of wider global change. It is based on very broad principles on which it is possible for a very wide range of viewpoints to agree.

Before concluding, a few clarifications may be needed. When we envisage people living their life in accordance with the central aim of humanity to reduce the distress of human beings and other forms of life, we do not have a vision either of ascetics or of people overburdened with difficult duties. Instead we envisage people who have time and space for song and dance, picnics and sightseeing and other numerous small joys of life. They are people who enjoy good food and see nothing wrong in wearing attractive clothes (although these should not be too expensive).

They voluntarily and happily accept a simple, frugal life as taking more than their share will be in conflict with the objective of caring for other human beings and other forms of life. They desire a deeper happiness and peace from their lifelong commitment to reducing the distress of other human beings and other forms of life. Due to this deeper commitment and the happiness this brings all the time, this work is done happily and so there is no feeling of being burdened by heavy duties.

Secondly, all aspects of this movement are entirely voluntary and peaceful.

The protective attitude towards other forms of life is emphasised, but of course it is taken for granted that people have the right to protect themselves from hostile forms of life, which may range from disease-spreading mosquitos to ferocious animals threatening children in a village.

Reducing Consumption
Probably also Increasing Happiness

IN MOST DISCUSSIONS of an alternative vision of society, it is generally agreed that the overwhelming importance accorded in the present day world to consumption and accumulation as the main source of satisfaction of human beings should be questioned and challenged. It is of course necessary to ensure that basic needs and a few small comforts are adequately met, it is also necessary to save for emergencies and old age (particularly in societies that do not yet provide medical insurance and old-age pensions).

But beyond these obvious needs, the relevance and desirability of consuming and accumulating, or making these the main focus of human endeavour - needs to be most certainly questioned and questioned seriously.

Among the millions of species who live on earth, humankind alone has the capacity to plan in an organised way for the protection and care of all forms of life as well as the natural systems that support these diverse forms of life. Hence it is clearly a tremendous waste of the wonderful potential of human beings to spend all or most of their efforts and energy in earning more and more dollars or euros.

This may appear simple and clear, but the fact remains that we've not been able to take this simple message to most people. The message of how one's basic needs having been adequately met, reduced emphasis on consumption and accumulation can actually lead to an increase in happiness needs to be explained more clearly to a larger audience. To explain this, moreover, we need persons who can themselves be seen and admired as living examples of this precept of 'reducing consumption, increasing happiness'.

Excessive consumption of many kinds – for example of the most coveted rich foods – can easily cause ill health. More generally, a life obsessed with a single-minded pursuit of money and material comforts can be extremely boring, pushing some to seek escape in several dangerous thrills, intoxications and misadventures.

As soon as we free ourselves from the never ending, boring and frequently destructive quest for more money and consumption, we begin to find more time and opportunity for a variety of experiences and interactions which are capable of bringing so much happiness in our everyday life. Our mind also opens up more, becomes more receptive to such experiences. Most of our experiences have to do with our relationships with our near and dear ones, but we also have opportunities to make new friends. We also have more time for - and discover our hidden capacity for – a variety of interesting and constructive skills which by themselves and by helping to enrich our everyday interactions add so much more meaning and delight to our life. For example, a rediscovery of our musical talents - long lost in early childhood experiences - can be richly rewarding in itself and also adds immensely to the quality of our interaction with our friends.

We also have more time to look at nature. And we suddenly discover the simple but great joys of watching a river flow, or closer to home, just absorbing closely the full beauty and majesty of a single flower or the amazing cuteness of a squirrel eating a nut. Or the affection of a sparrow feeding its baby. The song of a koel (nightingale), the dance of a peacock.

As we are drawn to the delightful gifts of mother nature, our protective instincts also grow. We begin to discover the happiness of protecting the forests in which birds can live happily. We learn that nothing can give greater happiness than protecting others, reducing the distress of others - whether of fellow human beings or other forms of life. This happiness, moreover, is a higher form of happiness, which no other form of happiness can equal.

Vegetarian Diet
The Reasons

MILLIONS OF SPECIES of mammals, birds, insects, amphibians and aquatic life exist on earth. Among them only humankind has the ability to significantly influence the welfare of other species, that too in a planned and coordinated way. Surely human beings should use this ability to provide as much protection as possible to other forms of life. The special strength and unique position of humankind should be used as a protector, and certainly not a destroyer, of other forms of life.

However, this basic protective role cannot be fulfilled if human beings persist in eating other mammals and birds (or other animal species).

It is a matter of choice for human beings whether they will confine themselves to plant-based food or else will consume animal-based food. It is also a matter of choice whether within animal foods, they'll consume that food which involves the upkeep of animals (milk and milk-based products) or that which involves the slaughtering of animals (meat and meat based products).

This choice exists for all human beings. But anyone who agrees that the basic role of human beings is to protect all other species of life, should have no hesitation in making a choice that excludes those animal foods which involve the killing of animals for human food.

The first principle of a vegetarian is that human beings should have a protective relation with other animals, birds and other such species. The second principle of a vegetarian is that human beings should not eat any other animal, bird or other such species as we can't protect those whom we agree to eat and relish.

The appetite of human beings, in terms of processing food wastefully into a wide range of concotions is so immense that as long as human beings continue to eat other animals, animals will continue to face many dangers and immense cruelty. Therefore, saying no to all food which involves the killing of animals is a must for all vegetarians.

In recent years there has been increasing recognition that vegetarian food is also healthier and also contributes to world's food security. These arguments are welcome, but these are not at all necessary to build a case against killing of animals for human food. For this just *one argument is enough – human beings should have a protective relationship with other animals and for this it is essential not to eat other animals.* However,food security and health arguments can be more usefully pursued in other cases where the 'killing' aspect is not so clear – such as in the case of human consumption of eggs.

If a vegetarian is against the killing of any animals, birds etc. for her (or his) diet, then surely, she should be equally concerned about not wearing anything that involves the killing of any animals. So a vegetarian also says no to any clothes which involve the killing of any animals for obtaining that cloth. To extend this principle further, a vegetarian refuse to use any consumer products which involve the killing of any animals, birds etc.

In a world of vegetarians, the character of animal husbandry will change. Animals will be kept for obtaining milk, wool and other useful products they provide while living, but no animals will be kept for slaughtering or for obtaining their meat. This will change the entire relationship between them - a relationship based on meeting eachother's needs, not a relationship that is destined to end in slaughter. The work of various artisans based on the skins and other body – parts of the animals who die a natural death is of course allowed.

In a world of vegetarians, hunting is altogether forbidden. However, any human being certainly retains the right to self-defence whenever attacked or seriously threatened by an animal. The collection of the remains of animals who die a natural death in the forest is permitted, subject to the caution that this provision is not misused.

These principles of a vegetarian life make it much easier for us to realise the protective role of human beings vis-a-vis other animal species.

However, this can also create a livelihood crisis for butchers, workers of meat industry, fisher-folk, hunters, certain types of artisans. On the other hand, there will be more opportunities for farmers who grow vegetables, fruits, legumes and cereals, as well as for artisans and traders processing and selling these products. By shifting more of the processing to rural artisans - where nutritious qualities can be saved - in fact an increase of livelihood opportunities can be obtained. As more people shift to vegetarian habits special efforts should be made to provide alternative livelihood opportunities to those who are displaced from their previous livelihood.

In particular, traditional small-scale fisher folk need protection from any adverse impact on the livelihood caused by the spread of vegetarian habits.

There is absolutely no scope or justification for forcing any of the principles of vegetarians on anyone. It is a purely voluntary movement. So even if the movement of vegetarian thinking makes steady progress, the progress is likely to be only gradual, providing enough time for a shift in livelihood opportunities.

PEACE, JUSTICE, EQUALITY

Just an old man in a loin cloth in distant India. Yet when he died, humanity wept.

—Louis Fischer

It is unwise to be too sure of one's own wisdom. It is healthy to be reminded that the strongest might weaken and the wisest might err.

—Mahatma Gandhi

In Retrospection

Grassroot Changes Needed

IN THIS BOOK we started by examining the growing relevance of Gandhiji's ideas for some of the most critical challenges facing our world. However, such issues cannot be resolved only by top-level changes. A strong base needs to be created at the grassroots by changes in the life and thinking of ordinary people.

So, we looked at this also and added to this the inspiring examples of some individuals and social movements whose work has been in keeping with Gandhian ideals.

We believe in keeping the window open for critical viewpoints also. Several criticisms of Gandhiji's personality aspects like 'less than just' attitudes towards some other highly dedicated freedom fighters, wartime recruitment, views on some aspects of public attitudes towards Second World War, celibacy experiments etc. can be quite justified.

While being aware of this, we nevertheless feel that it is more constructive at the present juncture of history to focus more on Gandhiji's essential message of peace, non-violence, universe as family, peaceful resistance of injustice. Simple life of need avoiding greed and compassion in economic thinking as this essential message is very important for resolving the most critical problems of humanity.

The big learning from Mahatma Gandhi's life and work has been that he did not merely convey these messages but put them in action in a hundred meaningful ways from which we can still learn a lot.

As humanity grapples with very critical survival issues in the first few decades of the 21st century, both at the wider policy level as well at the level of grassroots actions, the essential messages of Gandhiji for justice and equality, peace and environment protection, if understood correctly and in the right spirit, will continue to play a very inspirational and helpful role.

ACKNOWLEDGEMENTS

My family for all the support – particularly my wife Madhu.
I am grateful to my daughter Reshma Bharti
for her help in writing the section dealing with the work of
Badshah Khan and Khudai Khidmatgars.
Kumar Gautam for his valuable help in finalising the book,
Kuldeep Kulshrestha for his many-sided help.
Ishwar Singh for helping in many small ways.
My publishers, Renu Kaul Verma, Papri Sri Raman
and Team Vitasta for their kindness and support
AND
Special thanks to all the editors of newspapers
and journals who published parts of this book.
I have benefited greatly in my life from discussions with
several eminent Gandhian thinkers and activists. I remain
indebted to them.

About the Author

Bharat Dogra is a journalist and author who has written extensively on environment, development and welfare issues for over four decades.

He is also a leading Peace Activist. His articles and reports on these issues have been published in leading newspapers and journals in India and abroad. He has received several prestigious awards for his writings in English and Hindi. His reports from several remote parts of the country on rural distress and struggles have been well received.

On the basis of his wide-ranging experience and research, Bharat Dogra has increasingly argued in recent years that the environmental crisis is now taking the form of nothing less than a survival crisis and the next few years will be crucial for resolving this catastrophe. Some of these concerns were captured in his previous book published by Vitasta: *Survival Crisis-Planet in Peril; People's Response the Only Way Forward*.

In the present book he has drawn on all this experience to bring out the high relevance of Gandhian ideas and struggles to the most important problems and issues of the 21st century.

At present, he is Convener of *Save the Earth Now* Campaign, with its SED Demand.

For more details see: www.bharatdogra.in